# NOAA Technical Report NESDIS 142-8

# Regional Climate Trends and Scenarios for the U.S. National Climate Assessment

## Part 8. Climate of the Pacific Islands

Washington, D.C.
January 2013

**U.S. DEPARTMENT OF COMMERCE**
**National Oceanic and Atmospheric Administration**
National Environmental Satellite, Data, and Information Service

## NOAA TECHNICAL REPORTS
National Environmental Satellite, Data, and Information Service

The National Environmental Satellite, Data, and Information Service (NESDIS) manages the Nation's civil Earth-observing satellite systems, as well as global national data bases for meteorology, oceanography, geophysics, and solar-terrestrial sciences. From these sources, it develops and disseminates environmental data and information products critical to the protection of life and property, national defense, the national economy, energy development and distribution, global food supplies, and the development of natural resources.

Publication in the NOAA Technical Report series does not preclude later publication in scientific journals in expanded or modified form. The NESDIS series of NOAA Technical Reports is a continuation of the former NESS and EDIS series of NOAA Technical Reports and the NESC and EDS series of Environmental Science Services Administration (ESSA) Technical Reports.

Copies of earlier reports may be available by contacting NESDIS Chief of Staff, NOAA/ NESDIS, 1335 East-West Highway, SSMC1, Silver Spring, MD 20910, (301) 713-3578.

NOAA Technical Report NESDIS 142-8

# Regional Climate Trends and Scenarios for the U.S. National Climate Assessment

## Part 8. Climate of the Pacific Islands

**Victoria W. Keener**
East-West Center and the Pacific Regional Integrated Sciences & Assessments
Honolulu, HI

**Kevin Hamilton**
University of Hawai'i, International Pacific Research Center
Honolulu, HI

**Scot K. Izuka**
USGS Pacific Islands Water Science Center
Honolulu, HI

**Kenneth E. Kunkel, Laura E. Stevens, and Liqiang Sun**
Cooperative Institute for Climate and Satellites (CICS)
North Carolina State University and NOAA's National Climatic Data Center (NCDC)
Asheville, NC

**U.S. DEPARTMENT OF COMMERCE**
Rebecca Blank, Acting Secretary

**National Oceanic and Atmospheric Administration**
Dr. Jane Lubchenco, Under Secretary of Commerce for Oceans and Atmosphere
and NOAA Administrator

**National Environmental Satellite, Data, and Information Service**
Mary Kicza, Assistant Administrator

# PREFACE

This document is one of series of regional climate descriptions designed to provide input that can be used in the development of the National Climate Assessment (NCA). As part of a sustained assessment approach, it is intended that these documents will be updated as new and well-vetted model results are available and as new climate scenario needs become clear. It is also hoped that these documents (and associated data and resources) are of direct benefit to decision makers and communities seeking to use this information in developing adaptation plans.

There are nine reports in this series, one each for eight regions defined by the NCA, and one for the contiguous U.S. The eight NCA regions are the Northeast, Southeast, Midwest, Great Plains, Northwest, Southwest, Alaska, and Hawai'i/Pacific Islands.

These documents include a description of the observed historical climate conditions for each region and a set of climate scenarios as plausible futures – these components are described in more detail below.

While the datasets and simulations in these regional climate documents are not, by themselves, new, (they have been previously published in various sources), these documents represent a more complete and targeted synthesis of historical and plausible future climate conditions around the specific regions of the NCA.

There are two components of these descriptions. One component is a description of the historical climate conditions in the region. The other component is a description of the climate conditions associated with two future pathways of greenhouse gas emissions.

## Historical Climate

The description of the historical climate conditions was based on an analysis of core climate data (the data sources are available and described in each document). However, to help understand, prioritize, and describe the importance and significance of different climate conditions, additional input was derived from climate experts in each region, some of whom are authors on these reports. In particular, input was sought from the NOAA Regional Climate Centers and from the American Association of State Climatologists. The historical climate conditions are meant to provide a perspective on what has been happening in each region and what types of extreme events have historically been noteworthy, to provide a context for assessment of future impacts.

## Future Scenarios

The future climate scenarios are intended to provide an internally consistent set of climate conditions that can serve as inputs to analyses of potential impacts of climate change. The scenarios are not intended as projections as there are no established probabilities for their future realization. They simply represent an internally consistent climate picture using certain assumptions about the future pathway of greenhouse gas emissions. By "consistent" we mean that the relationships among different climate variables and the spatial patterns of these variables are derived directly from the same set of climate model simulations and are therefore physically plausible.

These future climate scenarios are based on well-established sources of information. No new climate model simulations or downscaled data sets were produced for use in these regional climate reports.

The use of the climate scenario information should take into account the following considerations:

1. All of the maps of climate variables contain information related to statistical significance of changes and model agreement. This information is crucial to appropriate application of the information. Three types of conditions are illustrated in these maps:

   a. The first condition is where most or all of the models simulate statistically significant changes and agree on the direction (whether increasing or decreasing) of the change. If this condition is present, then analyses of future impacts and vulnerabilities can more confidently incorporate this direction of change. It should be noted that the models may still produce a significant range of magnitude associated with the change, so the manner of incorporating these results into decision models will still depend to a large degree on the risk tolerance of the impacted system.

   b. The second condition is where the most or all of the models simulate changes that are too small to be statistically significant. If this condition is present, then assessment of impacts should be conducted on the basis that the future conditions could represent a small change from present or could be similar to current conditions and that the normal year-to-year fluctuations in climate dominate over any underlying long-term changes.

   c. The third condition is where most or all of the models simulate statistically significant changes but do not agree on the direction of the change, i.e. a sizeable fraction of the models simulate increases while another sizeable fraction simulate decreases. If this condition is present, there is little basis for a definitive assessment of impacts, and, separate assessments of potential impacts under an increasing scenario and under a decreasing scenario would be most prudent.

2. The range of conditions produced in climate model simulations is quite large. Several figures and tables provide quantification for this range. Impacts assessments should consider not only the mean changes, but also the range of these changes.

3. Several graphics compare historical observed mean temperature and total precipitation with model simulations for the same historical period. These should be examined since they provide one basis for assessing confidence in the model simulated future changes in climate.

   a. Temperature Changes: Magnitude. In most regions, the model simulations of the past century simulate the magnitude of change in temperature from observations; the southeast region being an exception where the lack of century-scale observed warming is not simulated in any model.

   b. Temperature Changes: Rate. The *rate* of warming over the last 40 years is well simulated in all regions.

   c. Precipitation Changes: Magnitude. Model simulations of precipitation generally simulate the overall observed trend but the observed decade-to-decade variations are greater than the model observations.

In general, for impacts assessments, this information suggests that the model simulations of temperature conditions for these scenarios are likely reliable, but users of precipitation simulations may want to consider the likelihood of decadal-scale variations larger than simulated by the models. It should also be noted that accompanying these documents will be a web-based resource with downloadable graphics, metadata about each, and more information and links to the datasets and overall descriptions of the process.

# 1. INTRODUCTION

The Global Change Research Act of 1990[1] mandated that national assessments of climate change be prepared not less frequently than every four years. The last national assessment was published in 2009 (Karl et al. 2009). To meet the requirements of the act, the Third National Climate Assessment (NCA) report is now being prepared. The National Climate Assessment Development and Advisory Committee (NCADAC), a federal advisory committee established in the spring of 2011, will produce the report. The NCADAC Scenarios Working Group (SWG) developed a set of specifications with regard to scenarios to provide a uniform framework for the chapter authors of the NCA report.

This climate document was prepared to provide a resource for authors of the Third National Climate Assessment report, pertinent to the state of Hawai'i and the United States-Affiliated Pacific Islands (US-API); hereafter referred to collectively as the Pacific Islands. The specifications of the NCADAC SWG, along with anticipated needs for historical information, guided the choices of information included in this description of Pacific Islands' climate. While guided by these specifications, the material herein is solely the responsibility of the authors and usage of this material is at the discretion of the 2013 NCA report authors.

This document has two main sections: one on historical conditions and trends, and the other on future conditions as simulated by climate models. The historical section concentrates on temperature and precipitation, primarily based on analyses of data from the National Weather Service's (NWS) Cooperative Observer Network, which has been in operation since the late 19th century. Additional climate features are discussed based on the availability of information. The future simulations section is exclusively focused on temperature and precipitation.

With regard to the future, the NCADAC, at its May 20, 2011 meeting, decided that scenarios should be prepared to provide an overall context for assessment of impacts, adaptation, and mitigation, and to coordinate any additional modeling used in synthesizing or analyzing the literature. Scenario information for climate, sea-level change, changes in other environmental factors (such as land cover), and changes in socioeconomic conditions (such as population growth and migration) have been prepared. This document provides an overall description of the climate information.

In order to complete this document in time for use by the NCA report authors, it was necessary to restrict its scope in the following ways. Firstly, this document does not include a comprehensive description of all climate aspects of relevance and interest to a national assessment. We restricted our discussion to climate conditions for which data were readily available. Secondly, the choice of climate model simulations was also restricted to readily available sources. Lastly, the document does not provide a comprehensive analysis of climate model performance for historical climate conditions, although a few selected analyses are included.

The NCADAC directed the "use of simulations forced by the A2 emissions scenario as the primary basis for the high climate future and by the B1 emissions scenario as the primary basis for the low climate future for the 2013 report" for climate scenarios. These emissions scenarios were generated by the Intergovernmental Panel on Climate Change (IPCC) and are described in the IPCC Special Report on Emissions Scenarios (SRES) (IPCC 2000). These scenarios were selected because they

---

[1] http://thomas.loc.gov/cgi-bin/bdquery/z?d101:SN00169:|TOM:/bss/d101query.html

5

incorporate much of the range of potential future human impacts on the climate system and because there is a large body of literature that uses climate and other scenarios based on them to evaluate potential impacts and adaptation options. These scenarios represent different narrative storylines about possible future social, economic, technological, and demographic developments. These SRES scenarios have internally consistent relationships that were used to describe future pathways of greenhouse gas emissions. The A2 scenario "describes a very heterogeneous world. The underlying theme is self-reliance and preservation of local identities. Fertility patterns across regions converge very slowly, which results in continuously increasing global population. Economic development is primarily regionally oriented and per capita economic growth and technological change are more fragmented and slower than in the other storylines" (IPCC 2000). The B1 scenario describes "a convergent world with…global population that peaks in mid-century and declines thereafter…but with rapid changes in economic structures toward a service and information economy, with reductions in material intensity, and the introduction of clean and resource-efficient technologies. The emphasis is on global solutions to economic, social, and environmental sustainability, including improved equity, but without additional climate initiatives" (IPCC 2000).

The temporal changes of emissions under these two scenarios are illustrated in Fig. 1 (left panel). Emissions under the A2 scenario continually rise during the 21$^{st}$ century from about 40 gigatons (Gt) $CO_2$-equivalent per year in the year 2000 to about 140 Gt $CO_2$-equivalent per year by 2100. By contrast, under the B1 scenario, emissions rise from about 40 Gt $CO_2$-equivalent per year in the year 2000 to a maximum of slightly more than 50 Gt $CO_2$-equivalent per year by mid-century, then falling to less than 30 Gt $CO_2$-equivalent per year by 2100. Under both scenarios, $CO_2$ concentrations rise throughout the 21$^{st}$ century. However, under the A2 scenario, there is an acceleration in concentration trends, and by 2100 the estimated concentration is above 800 ppm. Under the B1 scenario, the rate of increase gradually slows and concentrations level off at about 500 ppm by 2100. An increase of 1 ppm is equivalent to about 8 Gt of $CO_2$. The increase in concentration is considerably smaller than the rate of emissions because a sizeable fraction of the emitted $CO_2$ is absorbed by the oceans.

The projected $CO_2$ concentrations are used to estimate the effects on the earth's radiative energy budget, and this is the key forcing input used in global climate model simulations of the future. These simulations provide the primary source of information about how the future climate could evolve in response to the changing composition of the earth's atmosphere. A large number of modeling groups performed simulations of the 21$^{st}$ century in support of the IPCC's Fourth Assessment Report (AR4), using these two scenarios. The associated changes in global mean temperature by the year 2100 (relative to the average temperature during the late 20$^{th}$ century) are about +6.5°F (3.6°C) under the A2 scenario and +3.2°F (1.8°C) under the B1 scenario with considerable variations among models (Fig. 1, right panel).

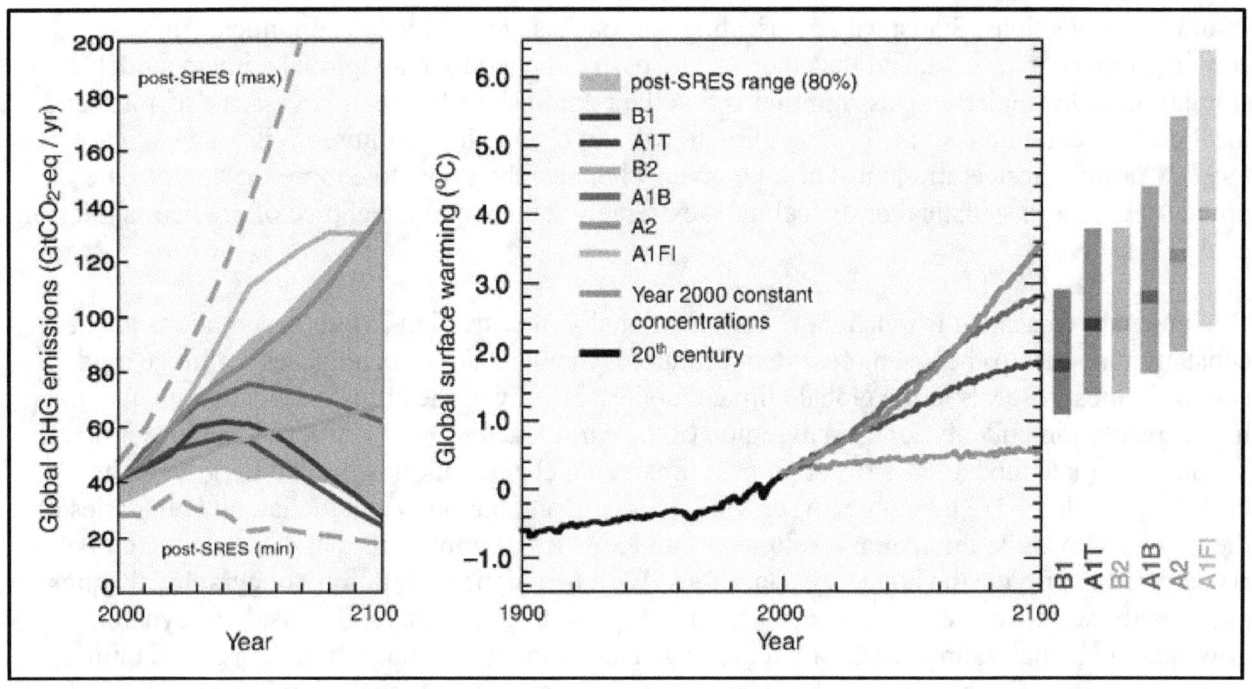

*Figure 1. Left Panel: Global GHG emissions (in GtCO$_2$-eq) in the absence of climate policies: six illustrative SRES marker scenarios (colored lines) and the 80$^{th}$ percentile range of recent scenarios published since SRES (post-SRES) (gray shaded area). Dashed lines show the full range of post-SRES scenarios. The emissions include CO$_2$, CH$_4$, N$_2$O and F-gases. Right Panel: Solid lines are multi-model global averages of surface warming for scenarios A2, A1B and B1, shown as continuations of the 20$^{th}$-century simulations. These projections also take into account emissions of short-lived GHGs and aerosols. The pink line is not a scenario, but is for Atmosphere-Ocean General Circulation Model (AOGCM) simulations where atmospheric concentrations are held constant at year 2000 values. The bars at the right of the figure indicate the best estimate (solid line within each bar) and the likely range assessed for the six SRES marker scenarios at 2090-2099. All temperatures are relative to the period 1980-1999. From IPCC AR4, Sections 3.1 and 3.2, Figures 3.1 and 3.2, IPCC (2007b).*

In addition to the direct output of the global climate model simulations, the NCADAC approved "the use of both statistically- and dynamically-downscaled data sets". "Downscaling" refers to the process of producing higher-resolution simulations of climate from the low-resolution outputs of the global models. The motivation for use of these types of data sets is the spatial resolution of global climate models. While the spatial resolution of available global climate model simulations varies widely, many models have resolutions in the range of 100-200 km (~60-120 miles). Such scales are very large compared to local and regional features important to many applications. For example, at these scales mountain ranges are not resolved sufficiently to provide a reasonably accurate representation of the sharp gradients in temperature, precipitation, and wind that typically exist in these areas.

Statistical downscaling achieves higher-resolution simulations through the development of statistical relationships between large-scale atmospheric features that are well-resolved by global models and the local climate conditions that are not well-resolved. The statistical relationships are developed by comparing observed local climate data with model simulations of the recent historical climate. These relationships are then applied to the simulations of the future to obtain local high-resolution projections. Statistical downscaling approaches are relatively economical from a computational perspective, and thus they can be easily applied to many global climate model simulations. One underlying assumption is that the relationships between large-scale features and local climate conditions in the present climate will not change in the future (Wilby and Wigley 1997). Careful consideration must also be given when deciding how to choose the appropriate predictors because statistical downscaling is extremely sensitive to the choice of predictors (Norton et al. 2011).

Dynamical downscaling is much more computationally intensive but avoids assumptions about constant relationships between present and future. Dynamical downscaling uses a climate model, similar in most respects to the global climate models. However, the climate model is run at a much higher resolution but only for a small region of the earth (such as North America) and is termed a "regional climate model (RCM)". A global climate model simulation is needed to provide the boundary conditions (e.g., temperature, wind, pressure, and humidity) on the lateral boundaries of the region. Typically, the spatial resolution of an RCM is 3 or more times higher than the global model used to provide the boundary conditions. With this higher resolution, topographic features and smaller-scale weather phenomena are better represented. The major downside of dynamical downscaling is that a simulation for a region can take as much computer time as a global climate model simulation for the entire globe. As a result, the availability of such simulations is limited, both in terms of global models used for boundary conditions and time periods of the simulations (Hayhoe 2010).

Section 3 of this document (Future Regional Climate Scenarios) responds to the NCADAC directives by incorporating analyses from multiple sources. The core source is the set of global climate model simulations performed for the IPCC AR4, also referred to as the Climate Model Intercomparison Project phase 3 (CMIP3) suite. These have undergone extensive evaluation and analysis by many research groups. A second source is a set of statistically-downscaled data sets based on the CMIP3 simulations. A third source is a set of dynamically-downscaled simulations, driven by CMIP3 models. A new set of global climate model simulations is being generated for the IPCC Fifth Assessment Report (AR5). This new set of simulations is referred to as the Climate Model Intercomparison Project phase 5 (CMIP5). These scenarios do not incorporate any CMIP5 simulations as relatively few were available at the time the data analyses were initiated. As noted earlier, the information included in this document is primarily concentrated around analyses of temperature and precipitation. This is explicitly the case for the future scenarios sections; due in large part to the short time frame and limited resources, we capitalized on the work of other groups on future climate simulations, and these groups have devoted a greater effort to the analysis of temperature and precipitation than other surface climate variables.

Climate models have generally exhibited a high level of ability to simulate the large-scale circulation patterns of the atmosphere. These include the seasonal progression of the position of the jet stream and associated storm tracks, the overall patterns of temperature and precipitation, the occasional occurrence of droughts and extreme temperature events, and the influence of geography on climatic patterns. There are also important processes that are less successfully simulated by models, as noted by the following selected examples.

Climate model simulation of clouds is problematic. Probably the greatest uncertainty in model simulations arises from clouds and their interactions with radiative energy fluxes (Dufresne and Bony 2008). Uncertainties related to clouds are largely responsible for the substantial range of

8

global temperature change in response to specified greenhouse gas forcing (Randall et al. 2007). Climate model simulation of precipitation shows considerable sensitivities to cloud parameterization schemes (Arakawa 2004). Cloud parameterizations remain inadequate in current GCMs. Consequently, climate models have large biases in simulating precipitation, particularly in the tropics. Models typically simulate too much light precipitation and too little heavy precipitation in both the tropics and middle latitudes, creating potential biases when studying extreme events (Bader et al. 2008).

Climate models also have biases in simulation of some important climate modes of variability. The El Niño-Southern Oscillation (ENSO) is a prominent example. In some parts of the U.S., El Niño and La Niña events make important contributions to year-to-year variations in conditions. Climate models have difficulty capturing the correct phase locking between the annual cycle and ENSO (AchutaRao and Sperber 2002). Some climate models also fail to represent the spatial and temporal structure of the El Niño - La Niña asymmetry (Monahan and Dai 2004). Climate simulations over the U.S. are affected adversely by these deficiencies in ENSO simulations.

The model biases listed above add additional layers of uncertainty to the information presented herein and should be kept in mind when using the climate information in this document.

The representation of the results of the suite of climate model simulations has been a subject of active discussion in the scientific literature. In many recent assessments, including AR4, the results of climate model simulations have been shown as multi-model mean maps (e.g., Figs. 10.8 and 10.9 in Meehl et al. 2007). Such maps give equal weight to all models, which is thought to better represent the present-day climate than any single model (Overland et al. 2011). However, models do not represent the current climate with equal fidelity. Knutti (2010) raises several issues about the multi-model mean approach. These include: (a) some model parameterizations may be tuned to observations, which reduces the spread of the results and may lead to underestimation of the true uncertainty; (b) many models share code and expertise and thus are not independent, leading to a reduction in the true number of independent simulations of the future climate; (c) all models have some processes that are not accurately simulated, and thus a greater number of models does not necessarily lead to a better projection of the future; and (d) there is no consensus on how to define a metric of model fidelity, and this is likely to depend on the application. Despite these issues, there is no clear superior alternative to the multi-model mean map presentation for general use. Tebaldi et al. (2011) propose a method for incorporating information about model variability and consensus. This method is adopted here where data availability make it possible. In this method, multi-model mean values at a grid point are put into one of three categories: (1) models agree on the statistical significance of changes and the sign of the changes; (2) models agree that the changes are not statistically significant; and (3) models agree that the changes are statistically significant but disagree on the sign of the changes. The details on specifying the categories are included in Section 3.

# 2. REGIONAL CLIMATE TRENDS AND IMPORTANT CLIMATE FACTORS

## 2.1. General Description of the Pacific Islands Climate

The Pacific Islands region extends from about 155°W to 130°E longitude and about 15°S to 25°N (Fig. 2). Due to its low-latitude location in the tropics, this region experiences relatively small seasonal variation in the energy received from the sun. This translates into a small annual air temperature range: across the region, there is only a 2-6°F difference between the warmest and coolest months. In contrast, precipitation varies greatly by both season and location. While the islands of the region all have distinct wet and dry seasons that loosely correspond to winter and summer months, the timing, duration, and intensity of these depends on a variety of sub-regional factors.

### 2.1.1. Central North Pacific

The warmest month in Hawai'i is August, with an average temperature of about 78°F; the coldest, February, averages around 72°F. Major geographic variations in temperature are due to high elevation. In the winter, the peaks of Mauna Kea and Mauna Loa can be covered with snow, with temperatures as low as 5°F. At elevations below 1,000 feet, however, nighttime lows rarely fall below the 50's. Hawai'i also shows striking geographic variation in rainfall. Annual rainfall averages can exceed 300 inches along the windward slopes of mountains, while in leeward coastal areas and on the upper slopes of the highest mountains, annual rainfall averages less than 20 inches (WRCC 2012).

The weather of Hawai'i is shaped by its proximity to the North Pacific High, a semi-permanent high-pressure area centered on 30°– 40°N and 140°–150°W, and its associated northeast trade winds. Trade wind conditions dominate from April to October, during which time the winds blow 70% of the time at speeds ranging from 10-25 miles per hour. These winds are accompanied by frequent light-to-moderate showers, due to the cumulus clouds being advected from the subtropical high. Regions of maximum rainfall are usually located on northeast, windward slopes where orographic uplifting is most pronounced, while showers are suppressed on southwestern leeward slopes as the air warms and sinks (NOAA 2012a).

Increased rainfall and storminess characterizes the Hawaiian winter. "Kona" Storms can generate widespread heavy rain and wind that last for days, as well as intense local showers for several hours. Kona storms are cut-off lows in the upper level subtropical Westerlies that usually occur to the north of Hawaii, and are associated with surface lows (NOAA 2012a).

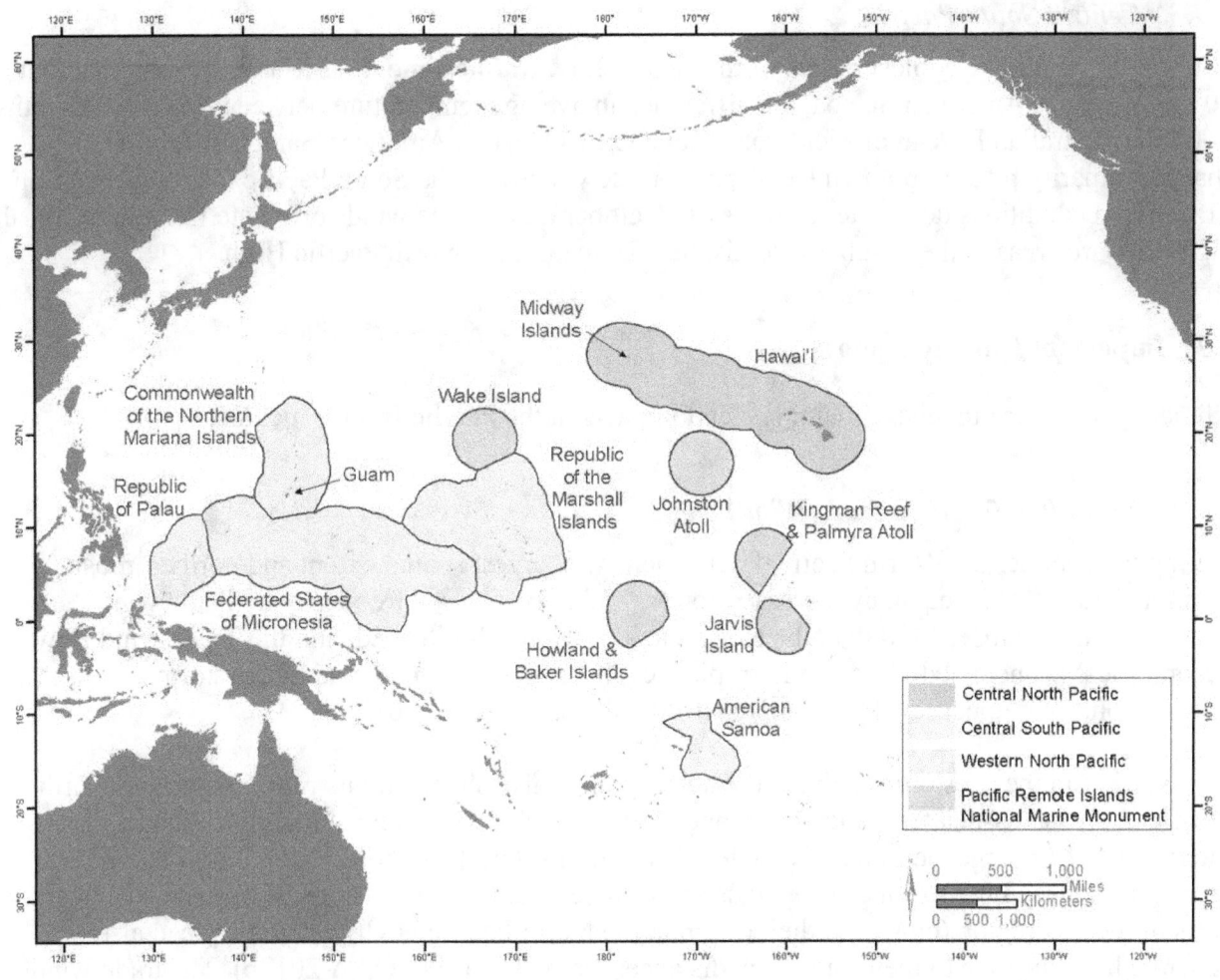

*Figure 2. Map of the Pacific Islands region and sub-regions. The region includes the Hawaiian archipelago and the US-Affiliated Pacific Islands, and comprises the Central North Pacific (CNP, blue), Western North Pacific (WNP, light orange), Central South Pacific (CSP, light green), and the islands of the Pacific Remote Island Marine National Monument (dark orange). Shaded areas indicate each island's exclusive economic zone (EEZ). From Keener et al. (2012).*

### 2.1.2. Western North Pacific

While warmest and coolest months can vary by location in the sub-region, there is a 5°F or less difference between the warmest and coolest months in each state. Similar to Hawaiʻi, geographic temperature variations can occur due to elevation, in that higher peaks of volcanic islands have lower temperatures. However, a different effect of elevation is felt by the smallest atoll islands in this sub-region, in that their temperature is affected strongly by sea surface temperature. For example, on Pohnpei the highest maximum and lowest minimum temperatures occur during the months of August-November. During the same months on Yap, though, minimum temperatures are consistent throughout the year, while maximum temperatures are highest in April-May (Australian Bureau of Meteorology and CSIRO 2011).

### 2.1.3. Central South Pacific

In this sub-region, the coolest month of the year is July, warmest month is March In Samoa, about 50 miles west of American Samoa, the difference in average temperature between these two months is 1.5°F (Australian Bureau of Meteorology and CSIRO 2011). American Samoa's weather is shaped primarily by the Southern Hemisphere trade winds and the South Pacific Convergence Zone. Trade wind conditions dominate from May to October, though the winds originate from subtropical high-pressure areas in the Southern Hemisphere instead of the North Pacific High.

## 2.2. Important Climate Factors

Climate phenomena that have major societal impacts in the Pacific Islands include:

### 2.2.1. North Pacific High/Trade Winds

During the summer, the North Pacific High reaches its largest spatial extent and northernmost position. Trade wind frequency increases correspondingly, as does frequency of Trade wind showers. From October to March, the North Pacific High is diminished, and the Trade wind flow becomes less pronounced. As a result, tropical cyclones, Kona storms, and major storm systems due to cold fronts and upper level lows approach the islands from the west.

The weather in the Western North Pacific sub-region is also shaped by its proximity to the North Pacific High and associated trade winds, and the monsoon trough (a local manifestation of the Intertropical Convergence Zone). Trade wind conditions dominate the Western North Pacific sub-region from December to June. The winds are stronger and last longer in the northern and eastern parts of the sub-region (closest to their origin in the North Pacific High) where they can also generate high surf due to their long fetch distances from Hawai'i (NOAA 2012b). The trade wind frequency is similar to that of Hawai'i (about 75% of the time). Unlike Hawai'i, though, the winds are more persistent during the winter months and less so during the summer months, when the monsoon trough expands, creating strong winds from the southwest (Kodama and Businger 1998). These gusty winds can last for multiple days, sometimes 1-2 weeks. The winds are more persistent in Palau and the southwest parts of the sub-region, episodic on Guam, and rarely felt in the Marshall Islands to the northeast (Bridgman and Oliver 2006; NOAA 2012b).

### 2.2.2. Intertropical Convergence Zone

The monsoon trough causing these trade winds is a stormy low-pressure system that arises due to differential heat absorption between the Asian continent and ocean. It is effectively a local, seasonal manifestation of the Intertropical Convergence Zone (ITCZ), which is a continual belt of low pressure near the equator. The position of the ITCZ varies seasonally, and from May to October, it moves through the Western North Pacific sub-region, bringing each island its rainy season. The northern parts of the sub-region (Guam and Mariana Islands) experience a dry season that is longer and drier, because of their proximity to the North Pacific High and trade winds. An active monsoon season strengthens the trough/ITCZ bringing additional rainfall to the western parts of the sub-region: Palau and the western islands in the Federated States of Micronesia (FSM) are most affected, while the Republic of the Marshall Islands (RMI) are only affected some years. The average annual rainfall across the sub-region ranges from 96 to 160 inches (Lander 2004; Australian Bureau of Meteorology and CSIRO 2011).

### 2.2.3. Tropical Cyclones

Monsoon trough activity is what makes the Western North Pacific sub-region the most active tropical cyclone basin, with an annual average of 25-26 cyclones reaching tropical storm strength or higher (Knapp, et al 2010) over the 30-year period from 1981-2010. These storms develop most commonly between July and December, and their frequency peaks in August, though they are possible throughout the year. Typhoons can lead to flooding, wind damage, coastal inundation, and reef erosion (NOAA 2012b).

There has been a shift in the occurrence of typhoons across the North West Pacific basin with far fewer storms occurring in the latter period from 1990-2010 compared to the period from1970-1990. This observation is quite consistent with the findings in other basins over this same time period. The increased proportion of major TCs in a basin with fewer overall TCs is quite consistent with findings by Webster et al. (2005), as well as by Maue (2011), and Diamond et al. (2012).

Tropical cyclones are also the primary form of extreme precipitation events in the Central South Pacific, where an average of 10-11 named storms occur based on a 41-year average from 1970-2010 (Diamond et al. 2012). Diamond et al. (2012) also noted that in this basin there were 83 major TCs that developed out of a total of 298 storms (27.9%) during the period 1970-1990, while from 1991-2010 there were 86 major TCs out of a total of 234 storms (36.7%). The increased proportion of major TCs over the past 20 years is statistically significant ($p < 0.005$, t-test with 19 degrees of freedom). Furthermore, of the 21 Category 5 storms that occurred in the region during that time, nearly 85% of them occurred during 1991-2010. The increased proportion of major TCs in a basin with fewer overall TCs is quite consistent with findings by Webster et al. (2005), as well as Maue (2011). For the entire period from 1970-2010, the area of greatest TC occurrence is west of the International Dateline in the areas around Vanuatu, New Caledonia, and Fiji.

### 2.2.4. South Pacific Convergence Zone

The South Pacific Convergence Zone (SPCZ) is a persistent band of cloudiness and storms, similar to the ITCZ, that originates in Southeast Asia and stretches southeast to French Polynesia. It is fueled by moisture from the warm western Pacific basin, and is most active from October to April (during the Southern Hemisphere summer) due to propagated monsoon activity from India (Vincent 1994).

Rainfall varies greatly by season in the Central South Pacific sub-region: approximately 75% of annual rainfall occurs from November to April, when the SPCZ is located about halfway between Western Samoa and Fiji. During the dry season, the SPCZ moves out of the area, and often becomes weak or inactive (Australian Bureau of Meteorology and CSIRO 2011).

### 2.2.5. El Niño-Southern Oscillation

The previous sections described the "normal" weather and climate across the Pacific Islands region in terms of local effects of large-scale atmospheric processes involving the North Pacific High, the trade winds, the ITCZ and SPCZ, and the East Asian and Western Pacific monsoon system. However, the atmospheric processes in this region are never in exact equilibrium, so variability in their local effects is inevitable. The El Niño-Southern Oscillation (ENSO) is an example of this variability on an interannual scale.

ENSO is a natural cycle of the climate system, recurring every 3-7 years and resulting from coupled ocean-atmosphere interaction. The tropical Pacific is notable for the large interannual variability of SST, surface winds, and rainfall connected with the ENSO phenomenon. The two extreme phases of this cycle are El Niño and La Niña. An El Niño phase is characterized by decreased trade wind activity, which allows the warm waters gathered by the winds into the west Pacific to flow eastward. This warm water, in turn, brings its associated cloudiness and rainfall. During the La Niña phase, this pattern is reversed: stronger trade winds allow the cold east Pacific to become larger. The effects of this cycle on the Pacific Islands region are significant, and help explain the large year-to-year variability in rainfall and other climatic variables (Chu 1995; Chu and Chen 2005; Giambelluca et al. 2011).

The large-scale ENSO variations have substantial effects on the interannual variations of rainfall seen in individual islands. Among the US-API, Yap, Palau, Chuuk, Guam, and the Northern Marianas Islands all share in the anomalously dry (wet) weather on average in the Western equatorial Pacific during the El Niño (La Niña) extremes of the Southern Oscillation. Hawai'i and Guam are located more centrally in the Pacific but also have observed correlations between seasonal rainfall and the state of ENSO, with El Niño generally being connected with anomalous dry weather in Hawai'i and wet weather in Samoa. The significant control of seasonal rainfall over these islands by ENSO suggests that long term changes in the Pacific basin-scale SST gradients and the Walker circulation will play a critical role in determining the mean rainfall changes on each of the islands. The Pacific Islands sub-regions are generally affected in the following ways:

### 2.2.5.1. Central North Pacific

Weakened trade wind during an El Niño event reduces rainfall, and causes dry conditions throughout the Hawaiian Islands (Chu and Chen 2005; Cao et al. 2007; Garza et al. 2012).

### 2.2.5.2. Western North Pacific

Weakened trade wind and loss of cloudiness/rainfall associated with warm water basin during an El Niño event leads to very dry conditions. Despite high annual rainfall totals, extended periods of dry conditions can endanger life. Frequency and intensity of subtropical cyclones (typhoons) are also affected by ENSO events (Lander 2004).

### 2.2.5.3. Central South Pacific

Location of the SPCZ is shifted during ENSO events, causing heavy rainfall or dry conditions (Australian Bureau of Meteorology and CSIRO 2011).

## 2.3. Climatic Trends

To fully assess the impact of climate variability and change and accurately predict future conditions in each sub-region, it is necessary to understand current and historic trends in climate and hydrologic records. For each sub-region, trends in observed data are discussed for three general types of records: temperature, rainfall, and extreme precipitation. Records such as temperature and rainfall are direct indicators of trends in climate. From these basic records, information on extreme precipitation events such as droughts or large storms can be extracted to provide important insight on how climate change can affect water resources. A common measure for drought is "consecutive dry days"; measures for extremely high precipitation include the frequency of high and moderate-intensity events, the frequency of typhoons and other storms, and the total rainfall over a specified number of consecutive rainy days.

### 2.3.1. Central North Pacific: Hawai'i

#### 2.3.1.1. Temperature

Generally, air temperature has increased significantly throughout the state of Hawai'i at both high and low elevations over the last century (Giambelluca et al. 2008). From 1919 to 2006, average temperature for stations in Hawai'i increased by 0.07°F per decade (Fig. 3). The rate of warming has accelerated to 0.11°F per decade in the last four decades. This statewide trend is only slightly lower than the global average trend of 0.13°F per decade from 1906 to 2005 (IPCC et al. 2007). The rate of increasing temperature is greater on high-elevation stations (0.13°F per decade at greater than 0.5 miles above sea level) (Fig. 3) and has been documented on the ecologically sensitive peaks of Haleakalā and Mauna Loa on Maui and Hawai'i Island, respectively, where the annual number of below-freezing days has decreased from 1958 to 2009 (Giambelluca et al. 2008; Diaz et al. 2011).

Much of the temperature variation prior to 1975 in Hawai'i appears to have been tightly coupled to the Pacific Decadal Oscillation (PDO) (Giambelluca et al. 2008). Since 1975, however, temperature in Hawai'i has diverged increasingly from the PDO (Fig. 4), which may indicate the increased influence of global warming (Giambelluca et al. 2008). Temperature data are consistent with an increase in the frequency of occurrence of the trade wind inversion (TWI) ) and a drop in trade-wind frequency over Hawai'i since the late 1970s (Fig. 5) (Cao et al. 2007; Garza et al. 2012), which is in turn consistent with continued warming and drying trends throughout Hawai'i, especially on high-elevation ecosystems.

#### 2.3.1.2. Precipitation

In Hawai'i, precipitation can manifest as rainfall, fog, hail, and snow. Annual precipitation over the state is variable, from 8 inches near the summit of Mauna Kea to over 400 inches on the windward slope of Haleakalā, Maui (Giambelluca et al. 2011). The dry summer season lasts from May to October, while the winter rainy season extends from November to April.

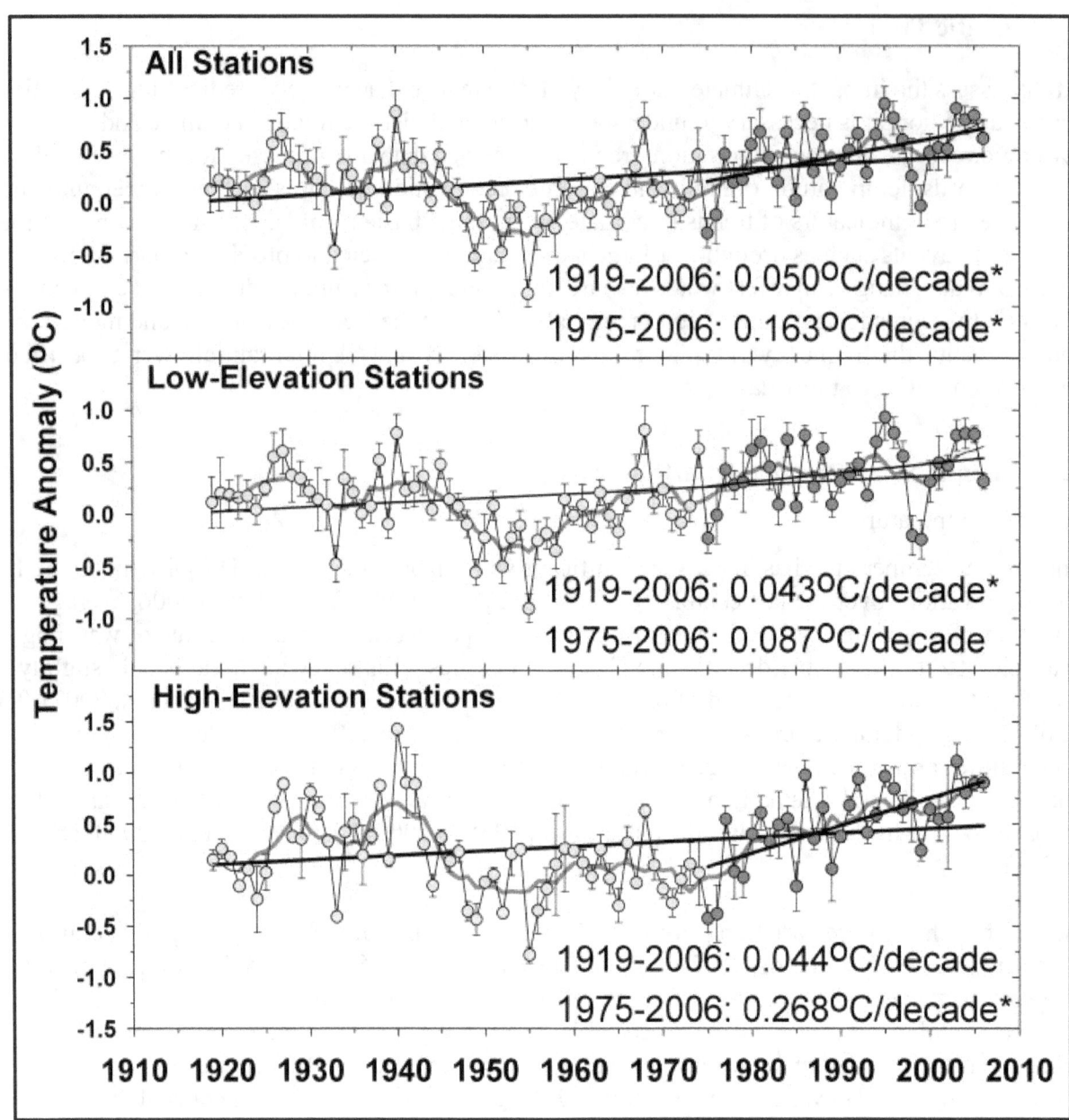

*Figure 3. Annual average surface temperature anomalies at stations in Hawai'i (a total of 21). Temperature anomalies are calculated first as the departure from the monthly mean, and then averaged into a calendar year. Anomalies greater than zero indicate temperatures that are above-average, while anomalies less than zero indicate below-average temperatures. A 7-year running-mean filter has been applied to the data to create a smoothed trend curve (black curve). Linear trends have been computed for two periods, 1919-2006 and 1975-2006, where the latter period emphasizes the observed enhanced warming. Temperature anomalies are increasing at both high and low-elevation stations. The steeper warming trend in high-elevation stations (>800 meters/0.5 mile) is visible in the bottom panel, especially when compared to that of the low-elevation stations in the middle panel (<800 meters/0.5 mile). Error bars show a standard deviation range of +/-0.5. Thick lines show 7-year running means. Asterisks indicate slopes significant at p = 0.05. Republished with permission of the American Geophysical Union, from Giambelluca et al. (2008); permission conveyed through Copyright Clearance Center, Inc.*

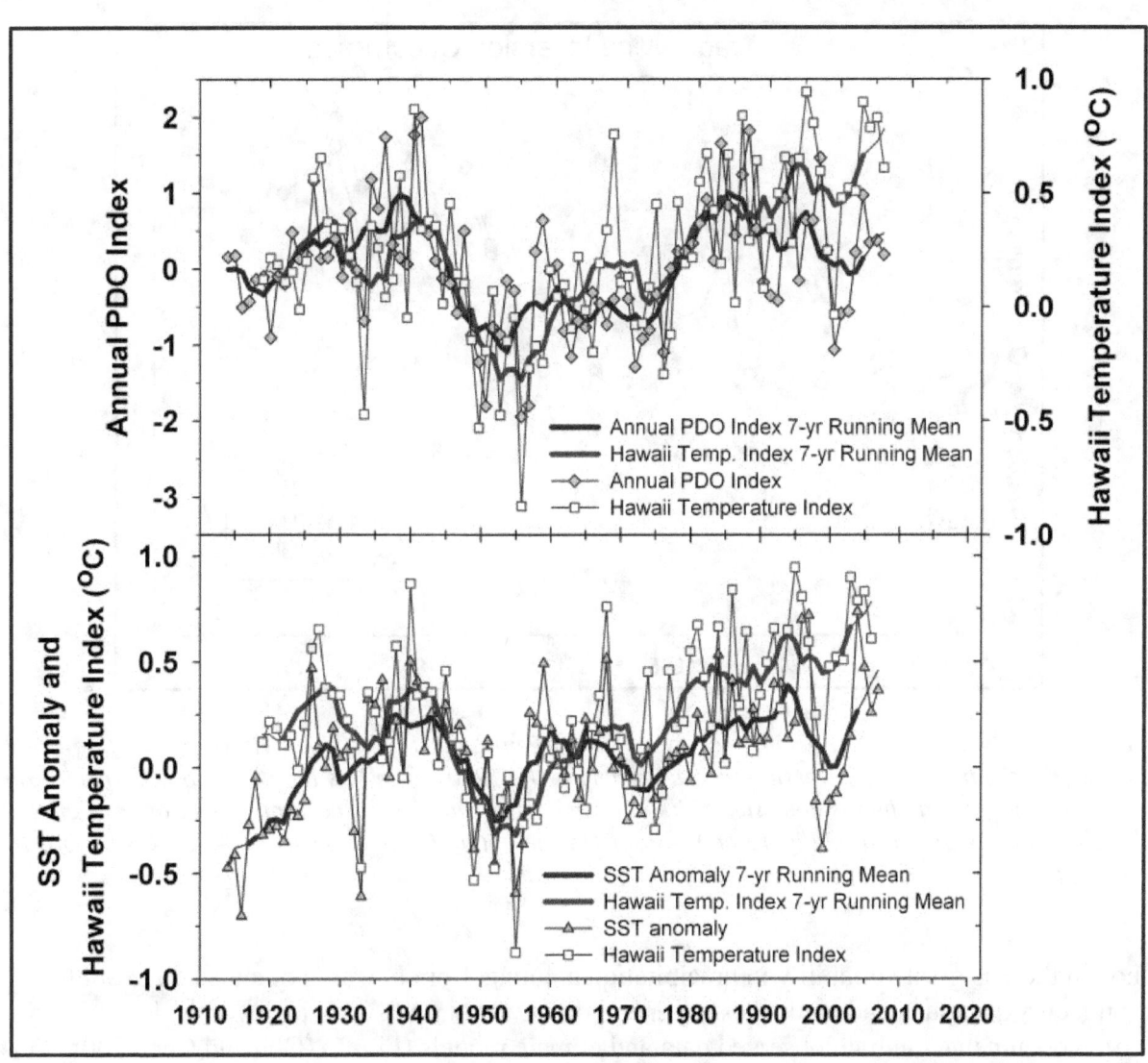

*Figure 4. The top panel shows the Hawai'i Temperature Index (red line) and the Pacific Decadal Oscillation (PDO) (blue line). Air Temperature prior to 1975 is tightly coupled to the PDO. Since 1975, air temperature has diverged increasingly from the observed PDO, which may indicate the increasing influence of climate change in the North Central Pacific sub-region. The bottom panel shows the local sea surface temperature anomalies (blue line) for 22°N, 156°W (based on the Smith and Reynolds (2004) Extended Reconstructed Sea Surface Temperatures (ERSST) data set), and the Hawai'i Temperature Index (red line). SST anomalies are also coupled to air temperatures, and show a similar decoupling around 1975. Republished with permission of the American Geophysical Union, from Giambelluca et al. (2008); permission conveyed through Copyright Clearance Center, Inc.*

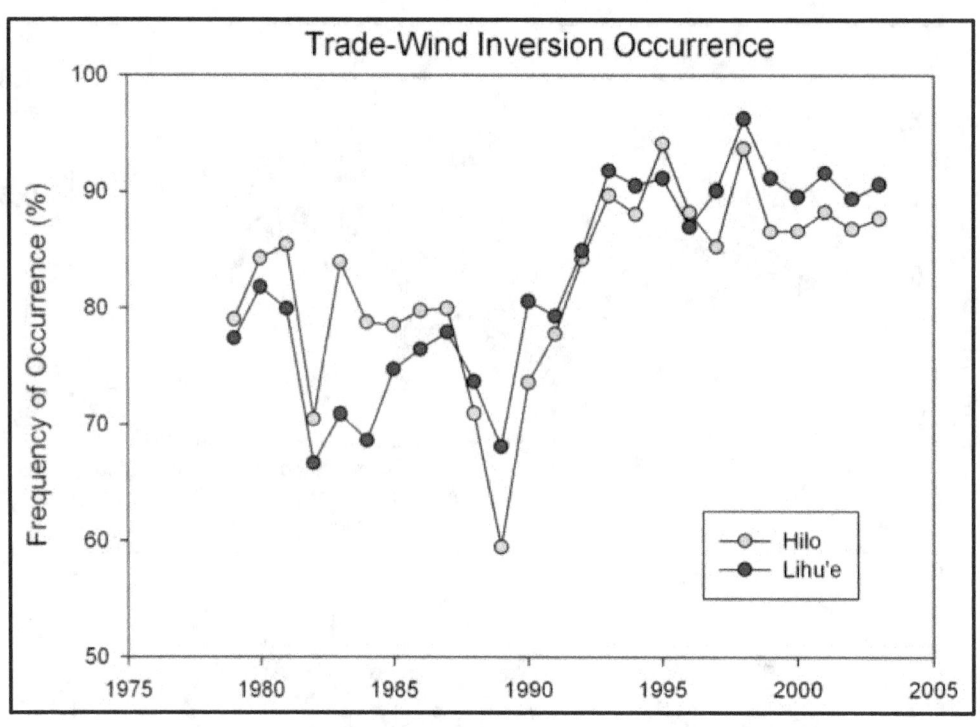

*Figure 5. Trade-wind inversion occurrence over Hilo (green) and Lihu'e (blue), Hawai'i. The frequency of occurrence of the trade-wind inversion increased during the 1990s from less than 80% to occurring around 90% of the time. This finding is consistent with current observations of reduced precipitation, especially for the high-elevations in Hawai'i. Figure courtesy of Hawai'i State Climatologist, Pao-Shin Chu. Data from Cao et al. (2007).*

Although there is great variability in precipitation amounts from one area of an island to another and on leeward versus windward sides, a general downward trend statewide over the last century has been documented in both observed data and climate models (Fig. 6) (Chu and Chen 2005; Diaz et al. 2011; Giambelluca et al. 2011; Timm et al. 2011). This decline in rainfall is consistent with an increase in the frequency of occurrence of the TWI, a decline in trade-wind occurrence, and corresponding higher rates of warming at high elevations (Cao et al. 2007; Diaz et al. 2011; Garza et al. 2012).

As with trends in air temperature, trends in precipitation in the wet season during the last century in Hawai'i were coupled tightly to the PDO until the mid-1970s. Precipitation variability in Hawai'i is also strongly affected by ENSO and the PDO. ENSO-scale patterns affect inter-annual variability whereas the PDO affects inter-decadal variability (Chu and Chen 2005). After air temperature and PDO diverged in 1975, initial evidence suggests that precipitation trends are following the same pattern of decoupling (Frazier et al. 2011) (Fig. 7). Climate change can affect ENSO and PDO patterns; this introduces greater uncertainty into future precipitation predictions for Hawai'i.

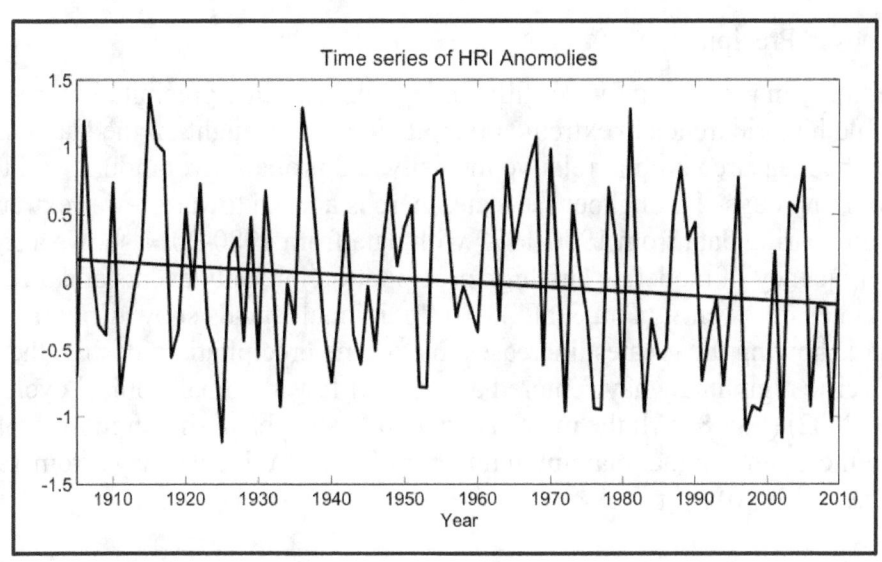

*Figure 6. Annual time series of the Hawai'i Rainfall Index (HRI) anomalies from 1905 to 2010. Annual values (July to June) are used. For example, 2010 refers to the period from July 2010 to June 2011. Nine stations from each of three islands (Kaua'i, O'ahu, and Hawai'i) are selected. These 27 stations are representative of the spatial variability of rainfall with regard to the direction of the prevailing northeast trade winds (i.e., windward and leeward) and to varying elevation levels (i.e., high and low). A normalization method is applied to each station to obtain standardized anomalies and the HRI is then calculated as the average of all station anomalies from three islands (Chu and Chen 2005). There is a long-term downward (drying) trend over the last 105 years (Keener et al. 2012).*

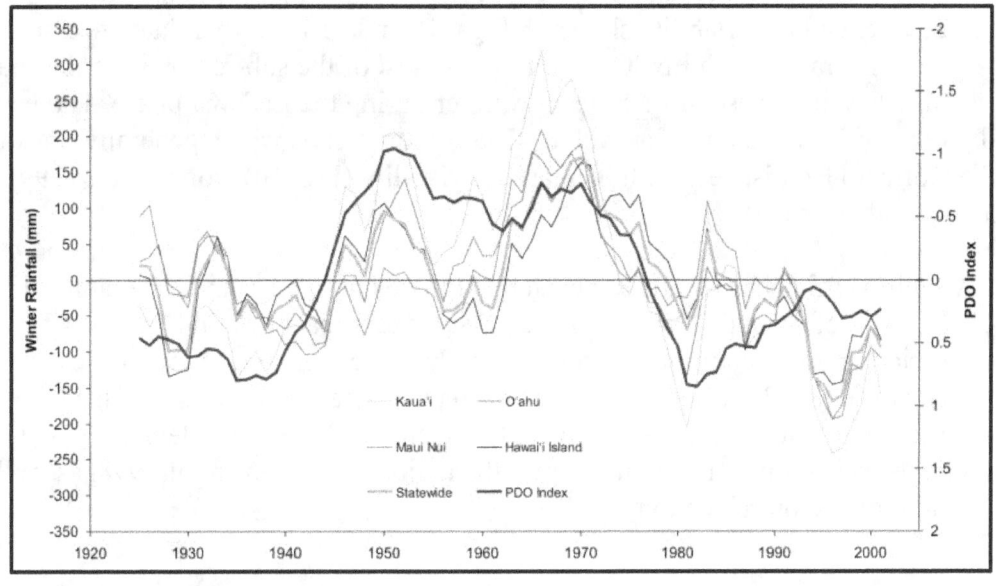

*Figure 7. Pacific Decadal Oscillation (PDO) Index, and precipitation anomalies at stations on four Hawaiian Islands in the wet season (November – April) that are smoothed decadally. Anomalies greater than zero indicate precipitation that is above-average, while anomalies less than zero indicate below-average precipitation. In trends similar to air temperature, precipitation anomalies have followed the same pattern of decoupling from the PDO since 1975, most likely as a result of long-term average climate change. Republished with permission of Elsevier, from Diaz and Giambelluca (2012); permission conveyed through Copyright Clearance Center, Inc.*

2.3.1.3.  Extremes in Precipitation

To reduce uncertainty in predicting future shifts in regional extreme precipitation, research has been done to understand historic trends in extreme precipitation and drought on the Hawaiian Islands. Precipitation can be measured both in relative intensity and probability, amount, and total amount over consecutive rainy days. Throughout the state, there is a trend toward fewer extremely high rainfall events: comparing data from 1950-1979 with data from 1980-2011 shows a significant decrease in the frequency of moderate and high precipitation intensity events and a corresponding increase in light intensity events, (Chu et al. 2010). Individual islands show different trends, with Oʻahu and Kauaʻi showing the greatest increases in extreme precipitation, despite the fact that the Central North Pacific region may have entered a period of fewer annual tropical cyclones since the mid-1990s (Chu 2002) (Fig. 8). All the major Hawaiian Islands have shown more prolonged drought, with an increasing annual maximum number of consecutive dry days from 1950-1979 to 1980-2011 (Chu et al. 2010) (Fig. 9).

### 2.3.2. Western North Pacific: West (Guam, Palau, FSM (Yap, Chuuk), CNMI); East (FSM (Pohnpei, Kosrae), RMI)

2.3.2.1.  Temperature

Many of the longest and most complete records of air temperature in the Western North Pacific sub-region are at airports and military air force bases. While there is at least one station on each major island group with a relatively complete and continuous record since 1950, other stations have very short records with many gaps. Across all recorded temperatures, however, observed maximum and minimum temperatures have exhibited visibly increasing trends over the past 60-years (Figs. 10 and 11) (Kruk and Marra 2012; Lander and Guard 2003; Lander and Khosrowpanah 2004; Lander 2004). The large interannual variability shown in Figs. 10 and 11 is partly related to a strong correlation of air temperature with ENSO conditions -- most of the sub-region is cooler than average during the El Niño phase of ENSO and warmer during the La Niña phase. The westernmost FSM island groups of Yap and Palau, as well as Guam show trends which generally track observed change in the Northern Hemisphere with increased variability (Fig. 10) (Jones et al. 1999; Brohan et al. 2006; Lander and Guard 2003).

The Majuro Weather Office has identified accelerated trends in maximum temperatures in the RMI since 1973, with a rise of around 0.25°F over the past 30 years (Jacklick and White 2011). In the same 30-year period, trends in minimum temperatures have been increasing more slowly, at about 0.22°F (Jacklick and White 2011). In both the western and eastern Micronesian island chains, the trends in increasing maximum temperatures have fallen since 2000. These declines may have to do with shifts in large-scale climate phenomena after the major El Niño event of 1998, as well as exhibiting artifacts of station relocation.

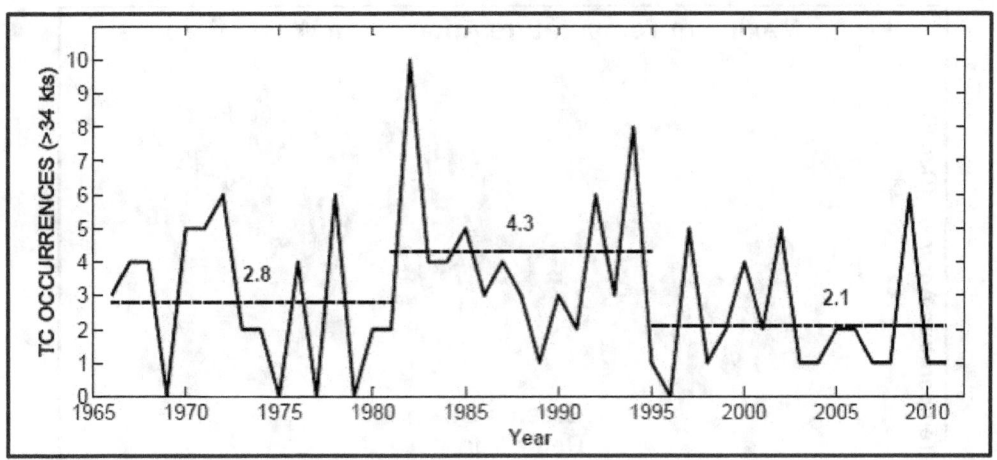

*Figure 8. Time series of tropical cyclones (tropical storms and hurricanes) in the Central North Pacific basin from 1966 to 2010. Although there was a period of greater tropical cyclone activity from the 1980s to mid-1990s, the basin has again entered a quieter time, with fewer average annual occurrences of storms. Broken lines denote the means for the periods 1966-81, 1982-94, and 1995-2010. Data from Chu (2002), figure from Keener et al. (2012).*

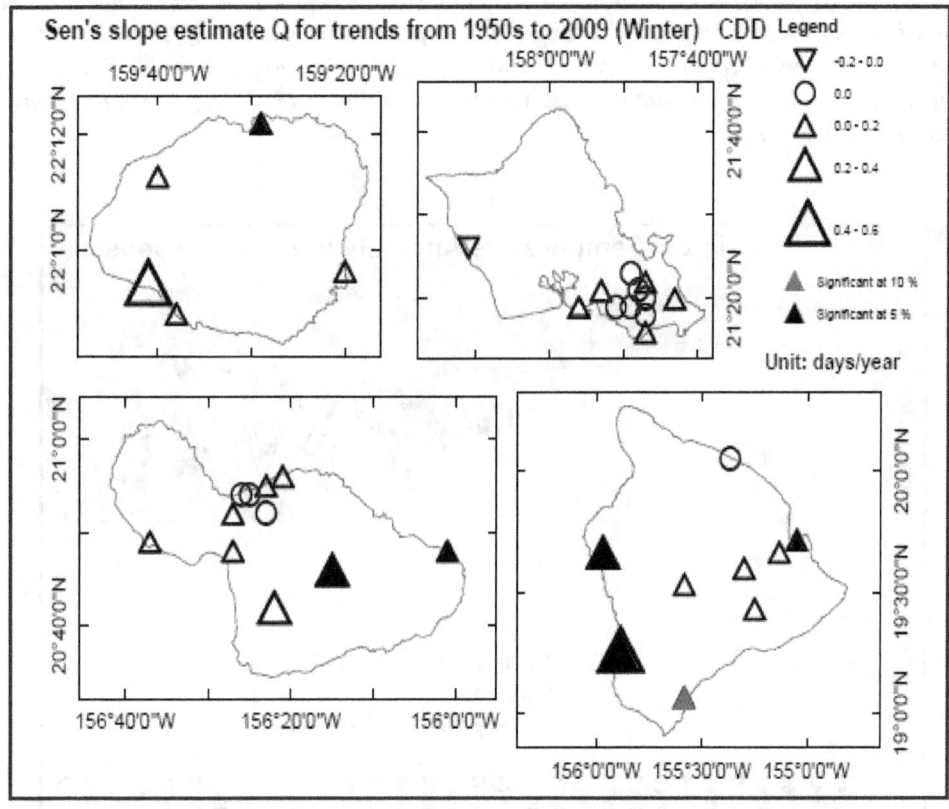

*Figure 9. Drought trends for Hawai'i. All four major Hawaiian Islands (O'ahu, Kaua'i, Maui and Hawai'i Island) have experienced increasing winter drought since the 1950s, defined by a longer annual maximum number of consecutive dry days. Upward (downward) triangles denote the increasing (decreasing) direction of drought trends. Black (gray) triangles indicate trends significant at the 5% (10%) level. Republished with permission of the American Meteorological Society, from Chu et al. (2010); permission conveyed through Copyright Clearance Center, Inc.*

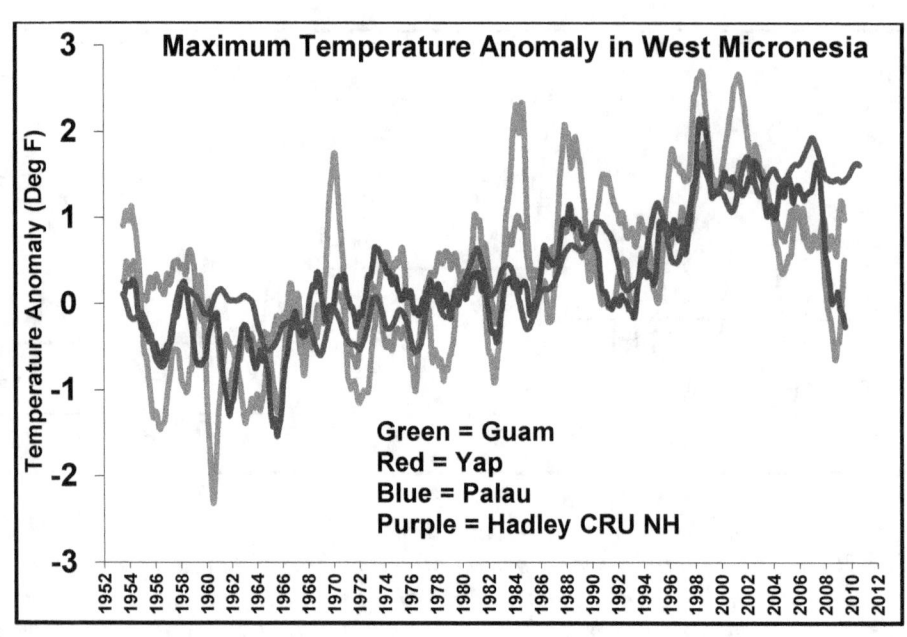

*Figure 10. Maximum monthly temperature anomaly time series from 1952 to 2012 for single monitoring stations with the most data in Yap, Guam, and Palau. The Northern Hemisphere temperature time series (purple line, Hadley CRU NH) is superimposed for comparison. Trends in maximum temperatures in the Western region of the Western North Pacific sub-region appear to be increasing at the same general rate as average Northern Hemisphere temperatures (Keener et al. 2012). Republished with <u>permission of the American Meteorological Society</u>, updated from Guard and Lander (2012); permission conveyed through Copyright Clearance Center, Inc.*

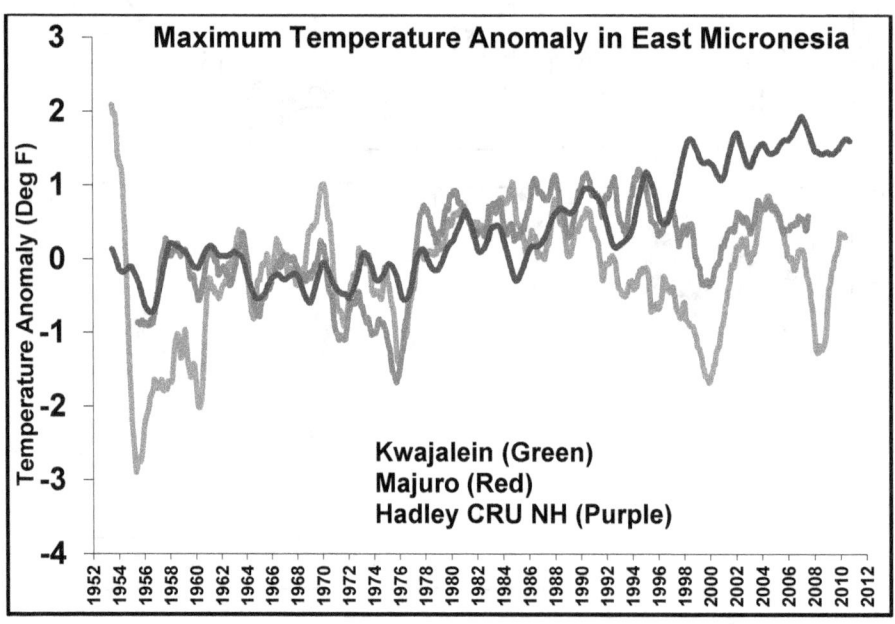

*Figure 11. Maximum monthly temperature anomaly time series from 1952 to 2012 for single monitoring stations with the most data in Kwajalein and Majuro in the RMI. The Northern Hemisphere temperature time series (purple line, Hadley CRU NH) is superimposed for comparison. Trends in maximum temperatures in the Eastern region of the Western North Pacific sub-region have high-levels of variability, and may reflect issues in the quality of the data or station infrastructure (Keener et al. 2012).*

## 2.3.2.2.  Precipitation

Islands throughout the Western North Pacific tend to receive abundant rainfall. Islands at lower latitudes such as Chuuk, Pohnpei, Kosrae and some atolls in the RMI receive over 118 inches of rainfall annually, which is stored in catchments as an important source of drinking and irrigation water (Lander and Khosrowpanah 2004; Bailey and Jenson 2011). There is a wet season and a dry season on all islands, with the relative length and intensity of each season depending on latitude (Yu et al. 1997). The more northward the placement of the island, the longer and drier the dry season tends to be. As with the variability in air temperature, ENSO has a strong effect on precipitation in Micronesia, with strong El Niño events corresponding closely with an increased risk for drought in the following year. Eastern islands in the sub-region such as Majuro and Kwajalein show a statistically significant drying trend from 1954 to 2011, such that over the last 60-years, these islands have lost nearly 15 percent of their annual rainfall, while western islands in the sub-region show a slight tendency towards wetter conditions (Fig. 12) (Bailey and Jenson 2011; Jacklick and White 2011). On the westernmost islands such as Palau and Yap, precipitation shows upward trends, but the trends are not statistically significant (Kruk and Marra 2012).


## 2.3.2.3.  Extremes in Precipitation

Although islands in the Western North Pacific sub-region have large amounts of rainfall annually, drought is a serious issue throughout Micronesia because of limited storage capacity and small groundwater supplies. Like Hawaii (Chu 1995), drought tends to be the most extreme during the winter and spring months following an El Niño. There is limited research on trends in extreme precipitation throughout Micronesia, although some results indicate that since the 1950s there are fewer extreme rainfall events greater than 10 inches in 24 hours in Guam and the Commonwealth of the Northern Mariana Islands (CNMI) (Lander and Guard 2003; Lander and Khosrowpanah 2004). Preliminary region-wide analysis indicates that both summer and winter 1-day amounts of precipitation over the 95th percentile have been declining since the early 1900s (Kruk and Marra 2012).


A more contentious issue for the sub-region is the trend in distribution and frequency of tropical cyclones (Knutson et al. 2010). The Western North Pacific Basin is the world's most prolific tropical cyclone basin, with an annual average of 25-26 named storms (Knapp et al 2010)[2]. Since 2000, the basin has been very quiet, with only 14 named storms in 2010 and 20 in 2011 (Camargo 2011 and Camargo 2012). While attribution of an individual tropical cyclone event to climate change is not possible, any upward shift in storm frequency in both the Western North Pacific basin and other basins around the world have destructive impacts on island nations, as typhoons tend to be more intense in El Niño years (Camargo and Sobel 2005), especially in the eastern portions of the sub-region. Research into how future climate will affect the frequency and intensity of tropical cyclones is of great societal importance for Pacific Islanders' food and water supply, livelihoods, and health (Gualdi et al. 2008; Murakami et al. 2011).

---

[2] This is based on a 30-year climatological average from 1981-2010.

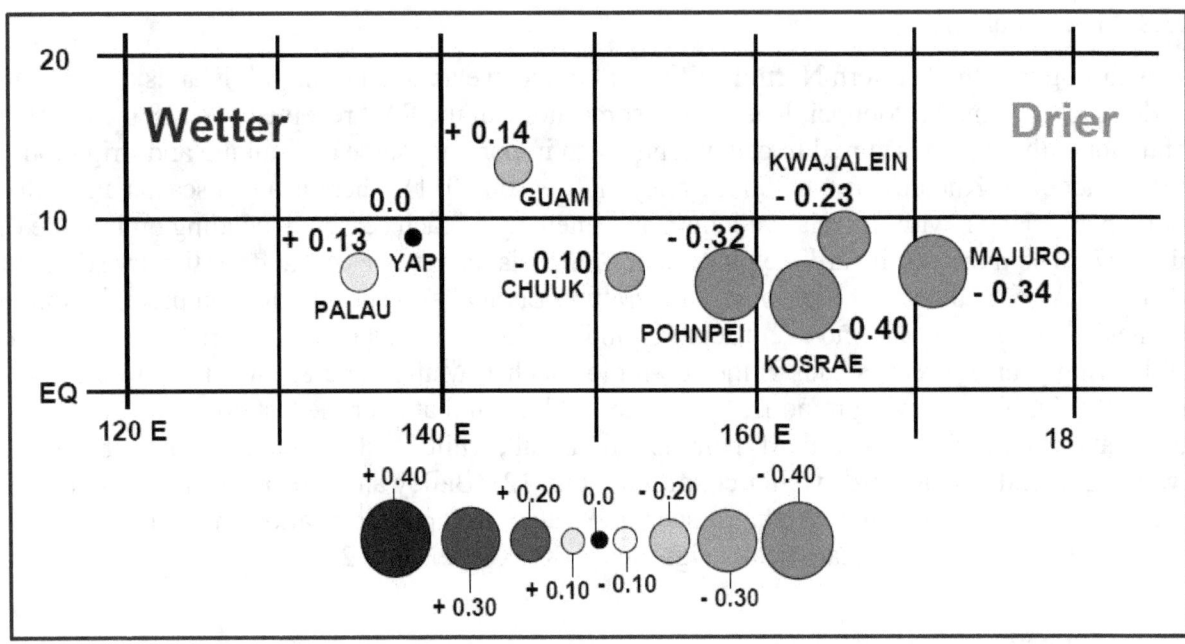

*Figure 12. Annual rainfall trends (inches per month per decade) in the Western North Pacific sub-region from 1950 to 2010. Blue shading indicates wetter, and red shading indicates drier. The size of the dot is proportional to the magnitude of the trend as per the inset scale. While islands in the West are tending towards getting slightly more precipitation, islands in the East are experiencing larger magnitudes of less precipitation (Keener et al. 2012).*

### 2.3.3. Central South Pacific: American Samoa

#### 2.3.3.1. Temperature

Average air temperatures in American Samoa are tropical, ranging from about 70-90°F. In the CSP sub-region, there is a general warming trend since the 1950s in average, minimum and maximum temperatures. The largest observed increase has been in minimum air temperatures, while average temperature increases range from 0.27°F to 0.45°F per decade, depending on the island (Australian Bureau of Meteorology and CSIRO 2011). Regional analyses of air temperature in Samoa are highly variable, but also show a rising trend in maximum air temperatures since 1950 (Young 2007).

#### 2.3.3.2. Precipitation

American Sāmoa is warm, humid, and rainy all year. The summer season is long and wet, lasting from October to May, and the winter season is only slightly cooler and drier, from June to September. Annual mean rainfall at Pago Pago Airport is about 3,048 mm (120 inches), although other areas can receive as little as 1,800 mm or as much as 5,000 mm (about 71 to 200 inches) due to orographic effect (Izuka et al. 2005). ENSO effects in American Sāmoa and the CSP vary by the strength of the particular anomaly event. During strong El Niño events, the monsoon trough is pulled northward and the SPCZ moves east-northeast of the Sāmoan region, making it significantly drier. In moderate El Niño events, the CSP is more susceptible to tropical cyclone formation and passage, and the rainy season tends to initiate earlier and end later. During weak El Niño events, the monsoon trough and SPCZ are west of the Samoa region. This causes reduced tropical storm activity and conditions that are drier than average. In Āpia, Sāmoa (about 80 km or 50 miles west of American Sāmoa), long-term records from 1890 to 2005 show no trend in daily, monthly, or annual precipitation (Young 2007; Australian Bureau of Meteorology & CSIRO 2011).

### 2.3.3.3. Extremes in Precipitation

Little detailed work has been undertaken examining trends in extreme events in this sub-region. Initial analysis of extreme precipitation records has only been done using the Pago Pago airport rain gauge, which has the longest period of record and the least missing data in American Sāmoa. Nearly all other rain gauges throughout American Sāmoa have been discontinued. Data from the Pago Pago Airport gauge show no trend in annual or winter one-day amounts of precipitation above the 95th percentile since 1965, and summer one-day amounts show a slight downward trend that is not statistically significant (Kruk and Marra 2012). ENSO and tropical cyclones are associated with extreme events in the South Pacific Islands. For the Central South Pacific sub-region, tropical cyclones occur between November and April; the number of cyclones varies widely from year to year but they tend to occur more frequently during moderate-intensity El Niño years and less frequently during weak El Niño events. Additionally, Madden-Julian Oscillation (MJO) propagation, the major source of intraseasonal variability in the tropical atmosphere, intensifies and increases the frequency of tropical cyclones during moderate El Niño events. Lastly, during strong La Niña events, the SPCZ lies far southwest of the Sāmoan region, and the risk of tropical cyclone development is moderate to high. The frequency of extremely high rainfall events per year has remained consistent since 1965 (Kruk and Marra 2012).

# 3. FUTURE REGIONAL CLIMATE SCENARIOS

As noted above, the physical climate framework for the 2013 NCA report is based on climate model simulations of the future using the high (A2) and low (B1) SRES emissions scenarios. The resulting climate conditions are to be viewed as scenarios, not forecasts, and there are no explicit or implicit assumptions about the probability of occurrence of either scenario.

## 3.1. Description of Data Sources

This summary of future regional climate scenarios is based on the global climate model output from phase 3 of the Coupled Model Intercomparison Project (CMIP3). Fifteen coupled Atmosphere-Ocean General Circulation Models (AOGCMs) from the World Climate Research Programme (WCRP) CMIP3 multi-model dataset (PCMDI 2012), as identified in the 2009 NCA report (Karl et al. 2009) were used: CCSM3, CGCM3.1 (T47), CNRM-CM3, CSIRO-Mk3.0, ECHAM5/MPI-OM, ECHO-G, GFDL-CM2.0, GFDL-CM2.1, INM-CM3.0, IPSL-CM4, MIROC3.2 (medres), MRI-CGCM2.3.2, PCM; UKMO-HadCM3, and UKMO-HadGEM1. The spatial resolution of the great majority of these model simulations was 2-3° (a grid point spacing of approximately 100-200 miles), with a few slightly greater or smaller. All model data were re-gridded to a common resolution before processing (see below). The simulations from all of these models include:

a) Simulations of the 20th century using best estimates of the temporal variations in external forcing factors (such as greenhouse gas concentrations, solar output, volcanic aerosol concentrations); and

b) Simulations of the 21st century assuming changing greenhouse gas concentrations following both the A2 and B1 emissions scenarios. The UKMO-HadGEM1 model did not have a B1 simulation.

CMIP3 multi-model mean analyses are provided for the periods of 2021-2050, 2041-2070, and 2070-2099, with changes calculated with respect to the historical climate reference period of 1971-2000. These future periods will be denoted in the text by their mid-points of 2035, 2055, and 2085, respectively. To produce the multi-model mean maps, each model's data is first re-gridded to a common grid of approximately 2.8° latitude (~190 miles) by 2.8° longitude (~130-170 miles). Then, each grid point value is calculated as the mean of all the available models' values at that grid point. Finally, the mean grid point values are mapped. This type of analysis weights all models equally. Although an equal weighting does not incorporate known differences among models in their fidelity in reproducing various climatic conditions, a number of research studies have found that the multi-model mean with equal weighting is superior to any single model in reproducing the present day climate.

## 3.2. Mean Temperature

Figure 13 shows the spatial distribution of multi-model mean simulated differences in average annual temperature for Hawai'i, for the three future time periods (2035, 2055, 2085) with respect to 1971-1999, for both emissions scenarios, for the 14 (B1) or 15 (A2) CMIP3 models. The statistical significance regarding the change in temperature between each future time period and the model reference period was determined using a 2-sample $t$-test assuming unequal variances for those two samples. For each period (present and future climate), the mean and standard deviation were calculated using the 29 or 30 annual values. These were then used to calculate $t$. In order to assess the agreement between models, the following three categories were determined for each grid point, similar to that described in Tebaldi et al. (2011):

- *Category 1*: If less than 50% of the models indicate a statistically significant change then the multi-model mean is shown in color. Model results are in general agreement that simulated changes are within historical variations;
- *Category 2*: If more than 50% of the models indicate a statistically significant change, and less than 67% of the significant models agree on the sign of the change, then the grid points are masked out, indicating that the models are in disagreement about the direction of change;
- *Category 3*: If more than 50% of the models indicate a statistically significant change, and more than 67% of the significant models agree on the sign of the change, then the multi-model mean is shown in color with hatching. Model results are in agreement that simulated changes are statistically significant and in a particular direction.

It can be seen from Fig. 13 that all three periods indicate an increase in temperature compared to 1971-1999, a continuation of the upward trend in mean temperature in the region over the last century. There is little or no spatial variation, especially for the A2 scenario. On a temporal scale, warming increases over time, and also increases between B1 and A2 for each respective period. For 2035, values for both emissions scenarios range between 1 and 2°F. For 2055, warming in B1 ranges over 1-3°F and for A2, ranges from 2 to 3°F. Increases by 2085 are larger still, with a 2-3°F range for B1 and a 4-5°F range for A2. The CMIP3 models indicate that temperature changes across Hawai'i, for all three future time periods and both emissions scenarios, are statistically significant. The models also agree on the sign of change, with all grid points satisfying category 3 above, i.e. the models are in agreement on temperature increases throughout the region for each future time period and scenario.

Global CMIP3 analyses from IPCC (2007a) indicate a robust pattern of warming over the Pacific between the late 20th century and late 21st century, which is larger near the equator and smaller in the subtropics.

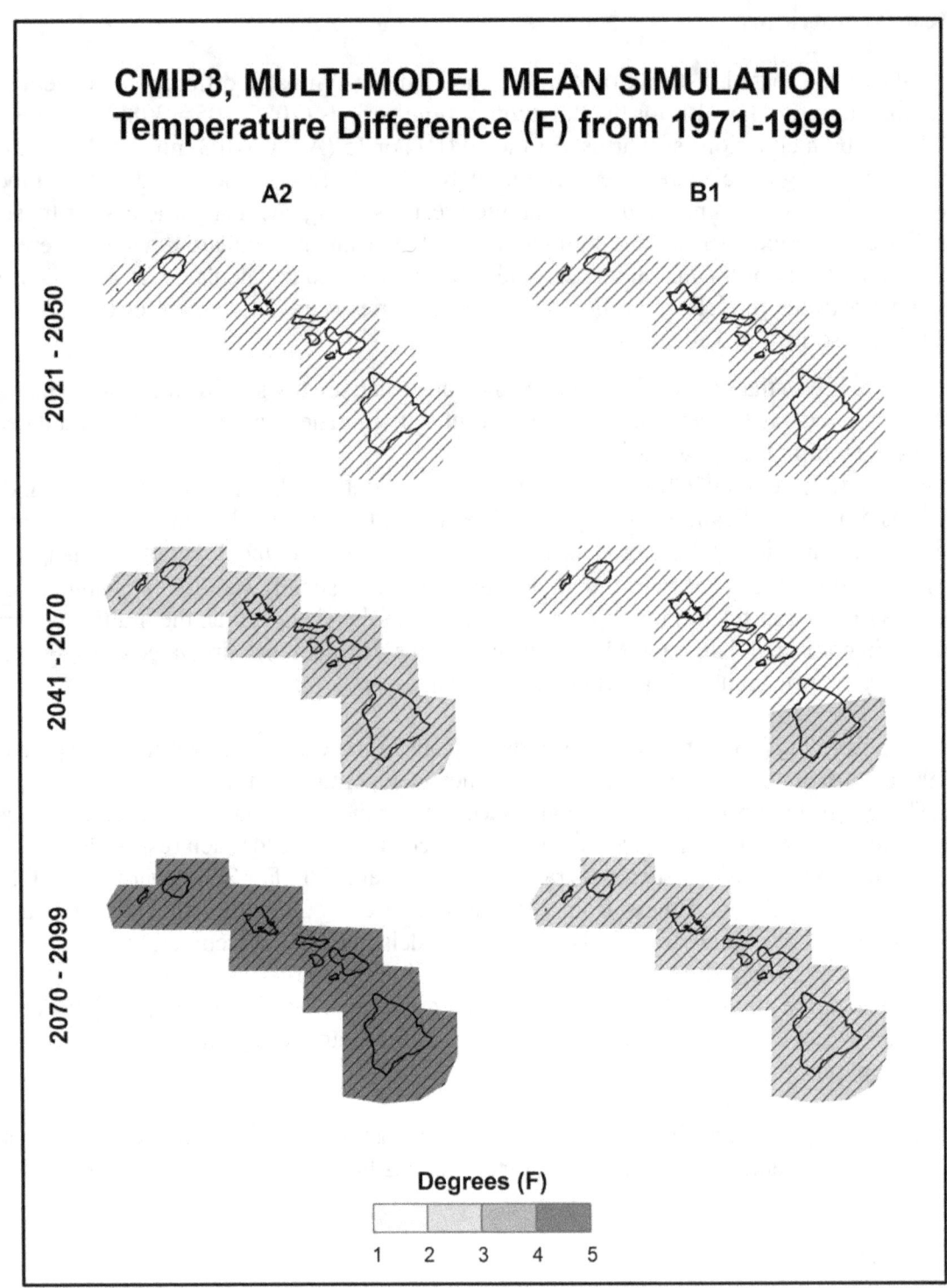

*Figure 13. Simulated difference in annual mean temperature (°F) for Hawai'i, for each future time period (2021-2050, 2041-2070, and 2070-2099) with respect to the reference period of 1971-1999. These are multi-model means for the high (A2) and low (B1) emissions scenarios from the 14 (B1) or 15 (A2) CMIP3 global climate simulations. Color with hatching (category 3) indicates that more than 50% of the models show a statistically significant change in temperature, and more than 67% agree on the sign of the change (see text).*

### 3.3. Mean Precipitation

The CMIP3 spatial distribution of multi-model mean simulated differences in average annual precipitation for Hawai'i is shown in Fig. 14, for the three future time periods (2035, 2055, 2085) with respect to 1971-1999, for both emissions scenarios, for the 14 (B1) or 15 (A2) CMIP3 models. Generally, there is a south-north gradient in changes. Despite a downward trend over the last century, by 2085 the southern parts of the region show relatively large increases while the northern areas show only slight decreases. This gradient increases in magnitude as time progresses for both scenarios The weakest spatial differences occur for the B1 scenario in 2035, with decreases of between 0 and 2% across the entire region. The agreement between models was once again assessed using the three categories described in Fig. 13. It can be seen that for all scenarios the changes in precipitation are not statistically significant for most models (category 1) over all grid points. This means that most models are in agreement that any changes will be smaller than the normal year-to-year variations that occur. For the A2 emissions scenario in 2085, the models are also in disagreement about the sign of the changes (category 2) over a large portion of the region.

Figure 15 shows some preliminary results analyzing the CMIP5 projections, in this case focusing on simulations with a single model (the GFDL CM3). Shown are time series of mean rainfall in the wet season each year averaged over a tropical Western Pacific region (June-September) and over a region around Hawai'i (December-March). Results are from a single realization forced with observed greenhouse gas concentrations from 1860 to 2005 and with projected concentrations using the CMIP5 RCP8.5 scenario through 2100. Considerable variability on interannual to interdecadal timescales is evident in both the Western Pacific and Hawaiian regions. However a clear tendency for increased wet season rainfall in the West Pacific as the climate warms is apparent even from this single realization, particularly in the projection for later in the present century.

Previous CMIP3 multi-model analyses from IPCC (2007a) show increases in precipitation along the equator and decreases over much of the subtropics, between the late 20th and 21st centuries. This basic pattern where "wet regions get wetter" under global warming has been explained as a result of increased moisture convergence due to higher absolute atmospheric humidity in the warmer climate (e.g., Held and Soden 2006), although Xie et al. (2010) show the importance also of changing spatial gradients in the SST in modifying the rainfall over the oceans at tropical and subtropical latitudes.

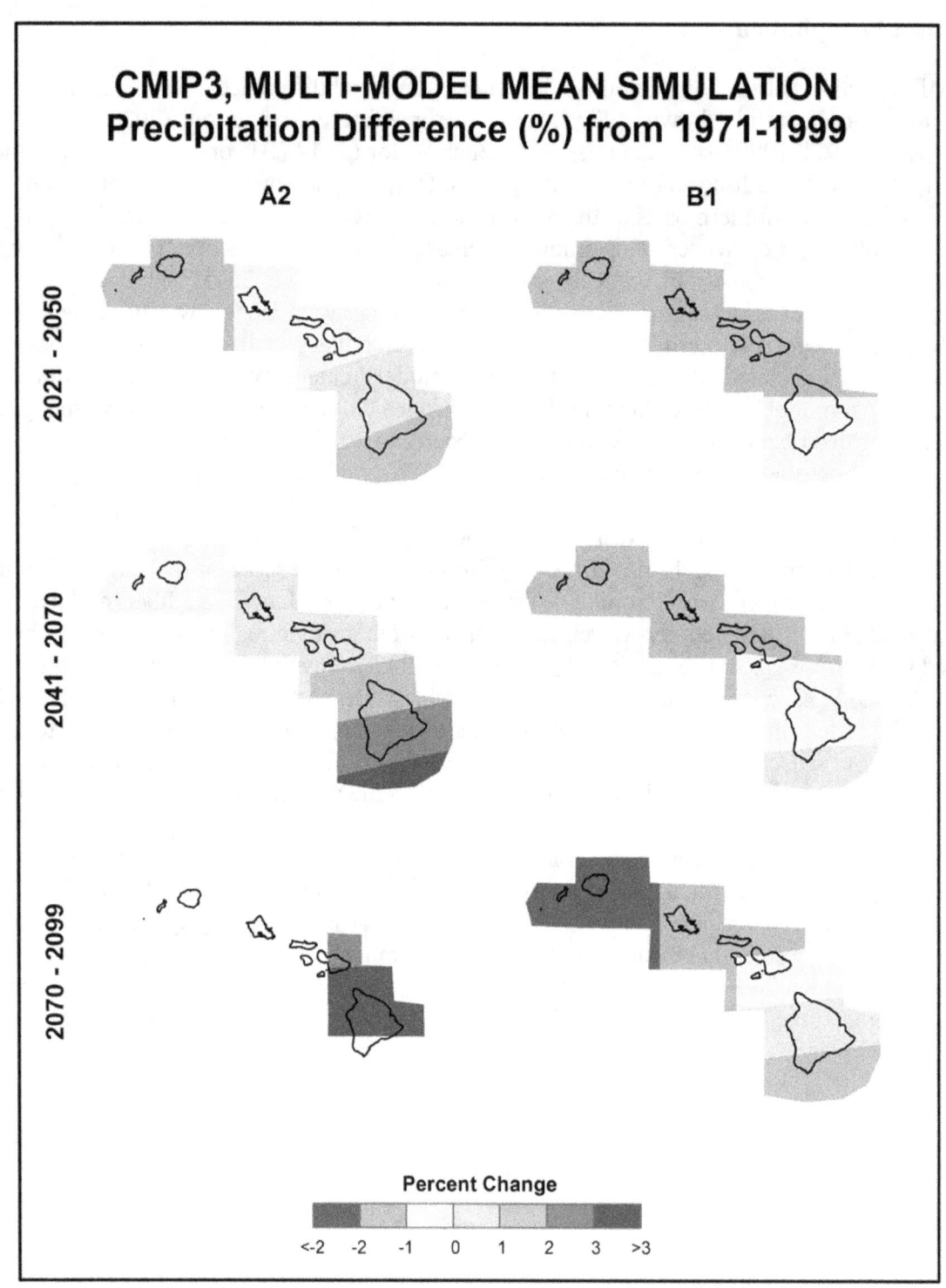

*Figure 14. Simulated difference in annual mean precipitation (%) for Hawai'i, for each future time period (2021-2050, 2041-2070, and 2070-2099) with respect to the reference period of 1971-1999. These are multi-model means for the high (A2) and low (B1) emissions scenarios from the 14 (B1) or 15 (A2) CMIP3 global climate simulations. Color only (category 1) indicates that less than 50% of the models show a statistically significant change in precipitation. Whited out areas (category 2) indicate that more than 50% of the models show a statistically significant change in precipitation, but less than 67% agree of the sign of the change (see text).*

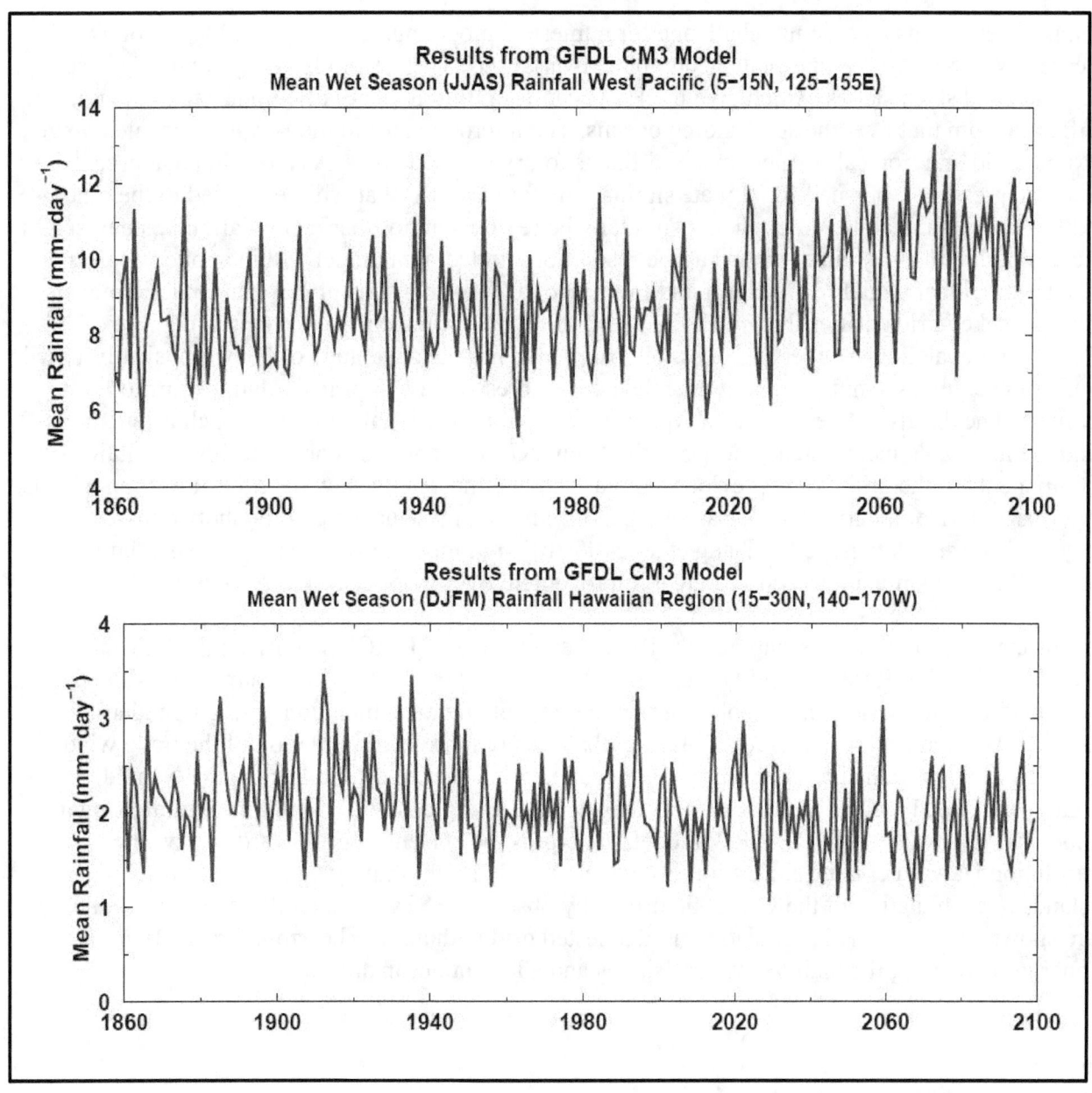

*Figure 15. Time series of mean rainfall in the wet season each year averaged over a tropical Western Pacific region (June-September, top) and over a region around Hawai‘i (December-March, bottom). Results are from a single realization forced with observed greenhouse gas concentrations from 1860 to 2005 and with projected concentrations using the CMIP5 RCP8.5 scenario through 2100. Figure courtesy of H. Annamalai and M. Mehari, based on data from the World Climate Research Programme's Working Group on Coupled Modelling CMIP5 project.*

Hawaiʻi represents a particular challenge for numerical modeling, and very fine resolution is necessary to resolve the fine-scale geographical variations. These include valleys and ridges, as well as broad and steep slopes, which give the Hawaiian Islands a diversity of climates that are quite different from that over the surrounding oceans. The microclimates in the Hawaiian Islands range from humid and tropical on the windward flanks to dry leeward areas. As a result, an approach now being aggressively pursued for climate simulations for Hawaiʻi (that will be applied to the other US-API as well) is the ability to explicitly simulate the regional atmospheric circulation at small scales. Such simulations may either embed a fine resolution limited-area model within a much coarser resolution global model (e.g., Giorgi and Francisco 2000) or may use global atmospheric models with stretched grids that enable much finer resolution over some particular region (e.g., Fox-Rabinovitz et al. 2006; Lal et al. 2008). This approach has the advantage of being physically-based and not needing assumptions about the relevance of present day empirical relationships to future climate. The disadvantages include a typically heavy computational burden for such calculations and the inevitable inconsistency between the simulated flows on the coarse and fine resolution components of the grid. Some previous simulations of atmospheric flow over the Hawaiian Islands have used horizontal grid spacings as small as 1.5 km, but these have been for short-term (a few days or less) simulations (e.g., Zhang et al. 2005a,b), or at most for seasonal forecasts (Nguyen et al. 2010). Such simulations showed great improvement over those at 10 km resolution.

A current project at the International Pacific Research Center (IPRC) is applying the Advanced Research Weather Research and Forecasting (WRF-ARW) model to climate simulations in Hawaiʻi, which will soon provide high-resolution regional model climate simulations using boundary conditions taken from CMIP3 and CMIP5 global simulations. The simulation of the trade wind boundary layer regime has been a particular challenge for numerical models (e.g., Wyant 2010; Zhang et al. 2011), but the recent success reported by Zhang et al. (2011) in this regard, using a modified version of the WRF-ARW model, underpins the current effort toward a very fine resolution Hawaiʻi climate simulation (Zhang et al. 2012). Preliminary results for rainfall over the islands in a simulation of the year 2006 driven by observed SSTs and lateral boundary conditions are shown in Fig. 16. This simulation used a nested grid with a 3-km horizontal resolution in the inner grid covering the main Hawaiian Islands and adjacent ocean areas.

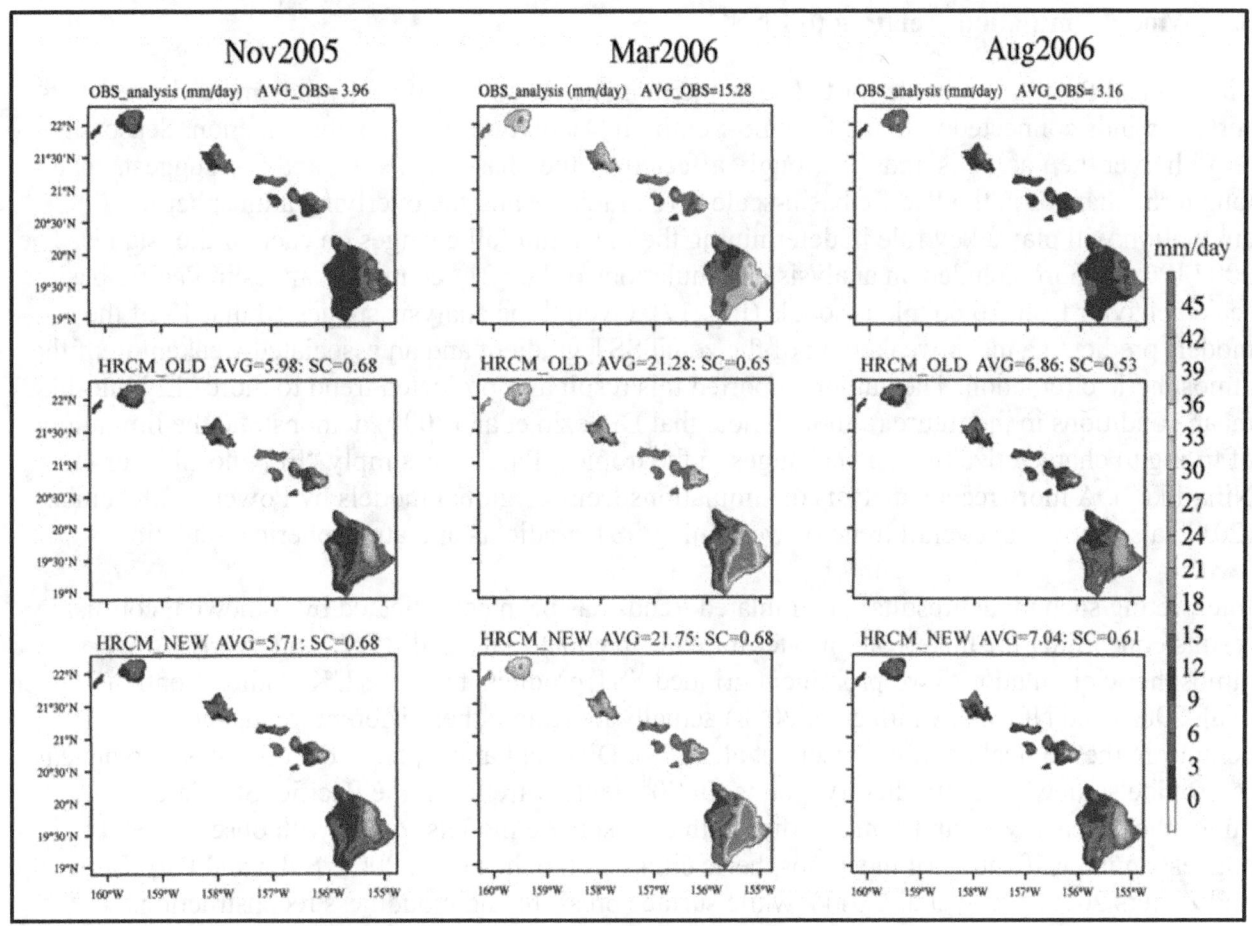

*Figure 16. Monthly rainfall over the main Hawaiian Islands from a one-year simulation with the regional WRF-ARW model, as described in Zhang et al. (2012). Results from two versions of the model are shown in the middle and lower rows. Observations in the top row are objective analyses based on rain gauge observations at a large number of stations. Results for November 2005, March 2006, and August 2006 are shown (Keener et al. 2012).*

## 3.4. Model simulations relating to ENSO

The tropical Pacific is notable for the large interannual variability of SST gradients and overlying surface winds connected with the El Niño-Southern Oscillation (ENSO) phenomenon. Seasonal rainfall over the Pacific islands is strongly affected by the phase of ENSO, and this suggests that long term changes in the Pacific basin-scale SST gradients and the overlying atmospheric circulation will play a key role in determining the mean rainfall changes on each of the islands. The 2007 IPCC report included an analysis of simulations of late 21$^{st}$ century changes in Pacific basin-scale behavior from 16 coupled models (Fig. 17). Overall the analysis suggested that 13 of the models predicted a mean weakening of the zonal SST gradient and an associated weakening of the atmospheric circulation. The authors reported this result as a projected trend to more "El Niño-like" mean conditions in the future (although note that DiNezio et al. (2009) demonstrate the limitations of trying to characterize the mean changes in the tropical Pacific as simply "El Niño-like" or "La Niña-like"). A more recent analysis of simulations from 21 global models by Power and Kociuba (2011) also shows an overall trend of weakening SST gradients and atmospheric circulation.

Interpreting such model results for simulated trends has been complicated by somewhat confusing results concerning the historical long-term trends in equatorial Pacific SST and overlying atmospheric circulation. Two prominent gridded SST products from the UK Hadley Centre (Rayner et al. 2003) and NOAA (Smith et al. 2008) actually indicate rather different trends over the 20$^{th}$ century in the tropical Pacific (Vecchi et al. 2008). Different atmospheric data sources also appear to provide somewhat contradictory results for 20$^{th}$ century trends in the Pacific. Standard atmospheric reanalyses and some studies with atmospheric models forced with observed SSTs suggest an intensification of the atmospheric circulation (Chen et al. 2008; Sohn and Park 2010; Yu and Zwiers 2010; Meng et al. 2011), while surface pressure and cloudiness reconstructions appear to be consistent with a modest weakening of the atmospheric circulation (Deser et al. 2010; Power and Kociuba 2011).

Recently the situation has improved with an increased understanding that the available historical data sets are affected by spurious trends introduced by changes in ship-based observations. Recently Tokinaga and Xie (2011) and Tokinaga et al. (2011) have made careful corrections for these trends, and have shown that the various observational data sets are indeed consistent with a slow weakening of the atmospheric circulation overlying much of the tropical Pacific during the 20$^{th}$ century. Certainly this is one aspect where progress has been made since the 2007 IPCC Report, and this new understanding will provide important context for modeling of climate trends in the Pacific region (Irving et al. 2011).

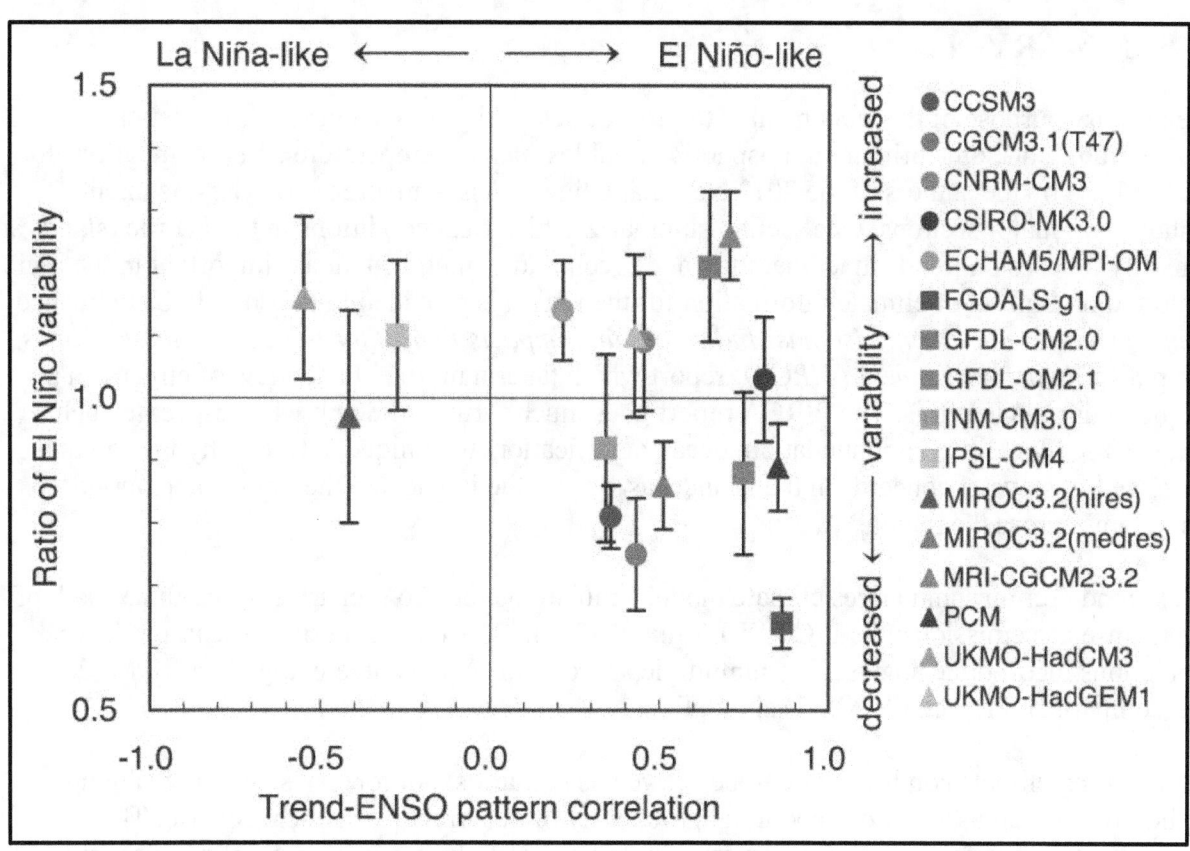

*Figure 17. Analysis of the simulated SST in the tropical Pacific in the late 21<sup>st</sup> century relative to that in the late 20<sup>th</sup> century in 16 global climate models. Changes in the amplitude of ENSO cyclic variations are plotted on the vertical axis, and a measure of changes in the mean SST gradients are indicated on the horizontal axis. The models show no pattern of agreement even on the sign of the change in strength of future ENSO oscillations, but most of the models agree in simulating a weakening of the mean gradients, which was interpreted roughly as a trend towards mean "El Niño-like" conditions. From IPCC AR4, Section 10.3.5.3, Figure 10.16, Meehl et al. (2007).*

# 4. SUMMARY

The primary purpose of this document is to provide selected physical climate information (concentrating on some primary atmospheric variables such as temperature and precipitation) for potential use by the authors of the 2013 National Climate Assessment report. The document contains two major sections. One section summarizes historical conditions in the Pacific Islands and primarily focuses on trends in temperature and precipitation metrics that are important in the region. A more thorough set of climate information for the region is provided in the soon to be published *Climate Change and Pacific Islands: Indicators and Impacts Report for the 2012 Pacific Islands Regional Climate Assessment (PIRCA)*[3] report which takes a more holistic view of climate in a unique region such as this. The PIRCA report goes much further in-depth with respect to such factors as sea-level rise and inundation, ocean acidification, the unique nature of hydrology and associated freshwater concerns in the islands, as well as the impacts to the unique and varied ecosystems across the region.

The second section summarizes climate model simulations for two scenarios of the future path of greenhouse gas emissions: the IPCC SRES high (A2) and low (B1) emissions scenarios. These simulations incorporate analyses from multiple sources, the core source being Coupled Model Intercomparison Project 3 (CMIP3) simulations.

The resulting climate conditions are to be viewed as scenarios, not forecasts, and there are no explicit or implicit assumptions about the probability of occurrence of either scenario. The basis for these climate scenarios (emissions scenarios and sources of climate information) were considered and approved by the National Climate Assessment Development and Advisory Committee.

Some key characteristics of the historical climate include:

- Climatic phenomena that have major impacts on the Pacific Islands include tropical cyclones, the El Niño-Southern Oscillation, drought, and variations in the position and strength of the trade winds and intertropical convergence zone.

- Average annual temperature has generally increased over the past 50-90 years. In Hawai'i, high elevation stations have been warming faster than low elevation stations over the past 30 years.

- There is a decline of northeast trade-wind frequency in Hawaii since 1973 and the frequency of trade-wind inversions in Hawai'i has increased over the past 20 years.

- Precipitation has trended downward over the past 100 years in Hawai'i. In the western Pacific, stations west of 150°E have become wetter while stations east of 150°E have been trending towards drier conditions.

- Hawai'i has experienced a trend toward increasing drought during the winter rainy season.

The climate characteristics simulated by climate models for the two emissions scenarios have the following key features:

---

[3] Report is available at http://www.cakex.org/NCAreports and http://www.islandpress.org/NCAreports.

- All three future time periods indicate an increase in temperature. Spatial variations are very small.

- Changes in precipitation are mixed. For Hawai'i, decreases in precipitation are simulated for the northern islands under the B1 scenario and during the early period under the A1 scenario. Increases are simulated for the southern islands under the A2 scenario. However, none of the changes are statistically significant.

- The occurrence of tropical cyclones across the Pacific is consistent with findings of decreased activity over the past 20 years, with indications that there is an increase in the occurrence of major storms over that same time period as well.

- Most models simulate a weakening of the east-west sea surface temperature gradient along the equator and an associated weakening of the wind patterns, which has been characterized as a shift to more El Niño-like conditions.

# 5. REFERENCES

AchutaRao, K., and K.R. Sperber, 2002: Simulation of the El Niño Southern Oscillation: Results from the Coupled Model Intercomparison Project. *Clim. Dyn.,* **19,** 191–209.

Arakawa, A., 2004: The cumulus parameterization problem: Past, present, and future. *J. Climate,* **17,** 2493-2525.

Australian Bureau of Meteorology and CSIRO, 2011: *Climate Change in the Pacific: Scientific Assessment and New Research,* CSIRO Publishing, 530 pp. [Available online at: http://www.cawcr.gov.au/projects/PCCSP/publications.html.]

Bader D. C., C. Covey, W. J. Gutowski Jr., I. M. Held, K. E. Kunkel, R. L. Miller, R. T. Tokmakian, and M. H. Zhang, 2008: *Climate models: An Assessment of Strengths and Limitations.* U.S. Climate Change Science Program Synthesis and Assessment Product 3.1. Department of Energy, Office of Biological and Environmental Research, 124 pp.

Bailey, R. T., and J. W. Jenson, 2011: Estimating the groundwater resources of atoll islands in the Federated States of Micronesia, Technical Report, Water and Environmental Research Institute of the Western Pacific, University of Guam, Guam.

Bridgman, H. A., and J. E. Oliver, 2006: *The Global Climate System: Patterns, Processes, and Teleconnections.* Cambridge University Press, 350 pp.

Brohan, P., J. Kennedy, I. Harris, S. Tett, and P. Jones, 2006: Uncertainty estimates in regional and global observed temperature changes: a new dataset from 1850, *J. Geophys. Res.,* **111,** D12106.

Camargo, S. J., 2011: The Tropics, Western North Pacific Basin. In: State of the Climate in 2010, J. Blunden, Arndt, D. S., and Baringer, M. O., Eds., *Bull. Am. Meteorol. Soc.,* 92, S123-S127.

——, 2012: The Tropics, Western North Pacific Basin. In: State of the Climate in 2011, J. Blunden, Arndt, D. S., and Baringer, M. O., Eds., *Bull. Am. Meteorol. Soc.,* 93, S107-S109.

Camargo, S. J., and A.H. Sobel, 2005: Western North Pacific tropical cyclone intensity and ENSO. *J. Climate,* **18,** 2996-3006.

Cao, G., T. W. Giambelluca, D. E. Stevens, and T. A. Schroeder, 2007: Inversion variability in the Hawaiian trade wind regime, *J. Climate,* **20,** 1145–1160.

Chen, J., A. D. Del Genio, B. E. Carlson, and M. G. Bosilovich, 2008: The spatiotemporal structure of twentieth-century climate variations in observations and reanalyses. Part I: Long-term trend, *J. Climate,* **21,** 2611–2633.

Chu, P.-S., 1995: Hawaii rainfall anomalies and El Niño. *J. Climate,* **8,** 1697-1703.

——, 2002: Large-Scale Circulation Features Associated with Decadal Variations of Tropical Cyclone Activity over the Central North Pacific, *J. Climate,* **15,** 2678.

Chu, P.-S., and H. Chen, 2005: Interannual and Interdecadal Rainfall Variations in the Hawaiian Islands, *J. Climate,* **18,** 4796-4813.

Chu, P.-S., Y. R. Chen, and T. A. Schroeder, 2010: Changes in precipitation extremes in the Hawaiian Islands in a warming climate, *J. Climate,* **23,** 4881–4900.

Deser, C., A. S. Phillips, and M. A. Alexander, 2010: Twentieth century tropical sea surface temperature trends revisited, *Geophys. Res. Lett.,* **37,** L10701.

Diamond, H. J., A. M. Lorrey, and J. A. Renwick, 2012: A Southwest Pacific tropical cyclone climatology and linkages to ENSO. *J. Climate*, in press. [Available online at http://journals.ametsoc.org/doi/abs/10.1175/JCLI-D-12-00077.1.]

Diaz H. F., and T. W. Giambelluca, 2012: Changes in atmospheric circulation patterns associated with high and low rainfall regimes in the Hawaiian Islands region on multiple time scales. *Global Planet. Change*, **98,** 97-108.

Diaz, H. F., P.-S. Chu, and J. K. Eischeid, 2005: Rainfall changes in Hawai'i during the last century. Preprint, *16th Conf. on Climate Variability and Change*, San Diego, CA. [Available online at: https://ams.confex.com/ams/pdfpapers/84210.pdf.]

Diaz, H. F., T. W. Giambelluca, and J. K. Eischeid, 2011: Changes in the vertical profiles of mean temperature and humidity in the Hawaiian Islands, *Global Planet. Change*. **77**, 21-25.

DiNezio, P. N., A. C. Clement, G. A. Vecchi, B. J. Soden, B. P. Kirtman, and S. K. Lee, 2009: Climate response of the equatorial Pacific to global warming, *J. Climate*, **22**, 4873-4892.

Dufresne, J.-L., and S. Bony. 2008: An assessment of the primary sources of spread of global warming estimates from coupled ocean–atmosphere models. *J. Climate,* **21**, 5135-5144.

Fox-Rabinovitz, M, B. Dugas, and J.L. McGregor, 2006: Variable resolution general circulation models: Stretched-grid model intercomparison project (SGMIP). *J. Geophys. Res.,* **111,** D16104.

Garza, JA., P.-S. Chu, C.W. Norton, and T.A. Schroeder, 2012: Changes of the prevailing trade winds over the islands of Hawaii and the North Pacific. *J. Geophys. Res.*, **117**, D11109.

Giambelluca, T. W., H. F. Diaz, and M. S. A. Luke, 2008: Secular temperature changes in Hawai'i, *Geophys. Res. Lett.*, **35,** L12702.

Giambelluca, T. W., Q. Chen, A. G. Frasier, J. P. Price, Y. L. Chen, P.-S. Chu, J. K. Eischeid, and D. M. Delparte, cited 2011: The Rainfall Atlas of Hawai'i. [Available online at: http://rainfall.geography.hawaii.edu.]

Giorgi F, and R. Francisco, 2000: Evaluating uncertainties in the prediction of regional climate change. *Geophys. Res. Lett.,* **27**, 1295-1298.

Gualdi, S., E. Scoccimarro, A. Navarra, and others, 2008: Changes in tropical cyclone activity due to global warming: Results from a high-resolution coupled general circulation model, *J. Climate*, **21**, 5204–5228.

Guard, C. P., and M. A. Lander, 2012: Northwest Pacific, Micronesia. In: State of the Climate in 2011, J. Blunden, Arndt, D. S., and Baringer, M. O., Eds., *Bull. Am. Meteorol. Soc.*, **93,** S215-S218.

Hayhoe, K. A., 2010: A standardized framework for evaluating the skill of regional climate downscaling techniques. Ph.D. thesis, University of Illinois, 153 pp. [Available online at https://www.ideals.illinois.edu/handle/2142/16044.]

Held, I. M., and B. J. Soden, 2006: Robust responses of the hydrological cycle to global warming. *J. Climate*, **19**, 5686-5699.

IPCC, 2000: *Special Report on Emissions Scenarios: A Special Report of Working Group III of the Intergovernmental Panel on Climate Change,* N. Nakicenovic, and R. Swart, Eds., Cambridge University Press, 570 pp.

——, 2007a: *Climate Change 2007: The Physical Science Basis. Contribution of Working Group I to the Fourth Assessment Report of the Intergovernmental Panel on Climate Change,* Solomon, S., D. Qin, M. Manning, Z. Chen, M. Marquis, K.B. Averyt, M. Tignor, and H.L. Miller, Eds., Cambridge University Press, 996 pp.

——, 2007b: *Climate Change 2007: Synthesis Report. Contribution of Working Groups I, II and III to the Fourth Assessment Report of the Intergovernmental Panel on Climate Change,* Pachauri, R. K, and Reisinger, A., Eds., IPCC, 104 pp.

——, cited 2012: IPCC Data Distribution Centre. [Available online at http://www.ipcc-data.org/ddc_co2.html.]

Irving, D. B., and Coauthors, 2011: Evaluating global climate models for the Pacific island region, *Clim. Res.,* **49,** 169–187.

Izuka, S. K., T. W. Giambelluca, and M. A. Nullet, 2005: Potential Evapotranspiration on Tutuila, American Samoa. U.S. Geological Survey Scientific Investigations Report 2005-5200, 40 pp. [Available online at: http://pubs.usgs.gov/sir/2005/5200/pdf/sir2005-5200.pdf.]

Jacklick, L. Z., and R. White, 2011: The Observed Climate, Climate Variability and Change of Majuro Atoll, Republic of the Marshall Islands. Pacific Climate Change Science Program, 1 pp. [Available online at: http://www.cawcr.gov.au/projects/PCCSP/pdf/5.Marshall_GH_Poster.pdf.]

Jones, P. D., M. New, D. E. Parker, S. Martin, and I. G. Rigor, 1999: Surface air temperature and its changes over the past 150 years, *Rev. Geophys.,* **37,** 173-200.

Karl, T. R., J. M. Melillo, and T. C. Peterson, Eds, 2009: *Global Climate Change Impacts in the United States.* Cambridge University Press, 188 pp.

Keener, V. W., J. J. Marra, M. L. Finucane, D. Spooner, and M. H. Smith, Eds., 2012: *Climate Change and Pacific Islands: Indicators and Impacts. Report for the 2012 Pacific Islands Regional Climate Assessment.* Island Press, 170 pp.

Knapp, K. R., M. C. Kruk, D. H. Levinson, H. J. Diamond, and C. J. Neumann, 2010: The International Best Track Archive for Climate Stewardship (IBTrACS): Unifying tropical cyclone best track data. *Bull. Am. Meteorol. Soc.,* 91, 363-376.

Knutson, T. R., J. L. McBride, J. Chan, K. Emanuel, G. Holland, C. Landsea, I. Held, J. P. Kossin, A. Srivastava, and M. Sugi, 2010: Tropical cyclones and climate change, *Nature Geoscience,* **3,** 157–163.

Knutti, R., 2010: The end of model democracy? *Climatic Change,* **102,** 395-404.

Kodama, K. R., and S. Businger, 1998: Weather and forecasting challenges in the Pacific Region of the National Weather Service. *Weather Forecast.,* **13**, 523-546.

Kruk, M. C., and J. J. Marra, 2012: A regional intercomparison of rainfall extremes. 92nd Annual Meeting of the Amer. Meteor. Soc., New Orleans, LA.

Lal, M., J. L. McGregor and K. C. Nguyen, 2008: Very high-resolution climate simulation over Fiji using a global variable-resolution model. *Clim. Dyn.,* **30**, 293-305.

Lander, M. A., 2004: Rainfall climatology for Saipan: Distribution, return-periods, El Niño, tropical cyclones, and long-term variations. Technical Report No. 103, Water and Environmental

Research Institute of the Western Pacific, University of Guam, 54 pp. [Available online at: http://www.weriguam.org/docs/reports/103.pdf.]

Lander, M. A., and C. P. Guard, 2003: Creation of a 50-year rainfall database, annual rainfall climatology, and annual rainfall distribution map for Guam. Technical Report No. 102, Water and Environmental Research Institute of the Western Pacific, University of Guam, 26 pp. [Available online at: http://www.weriguam.org/docs/reports/102.pdf.]

Lander, M. A., and S. Khosrowpanah, 2004: A rainfall climatology for Pohnpei Island the Federated States of Micronesia. Technical Report No. 100, Water and Environmental Research Institute of the Western Pacific, University of Guam, 51 pp. [Available online at: http://www.weriguam.org/docs/reports/100.pdf.]

Maue, R. N., 2011: Recent historically low global tropical cyclone activity. *Geophys. Res. Lett.*, **38**, L14803.

Meehl, G. A., and Coauthors, 2007: Global climate projections. *Climate Change 2007: The Physical Basis. Contribution of Working Group I to the Fourth Assessment Report of the Intergovernmental Panel on Climate Change*, Solomon, S., D. Qin, M. Manning, Z. Chen, M. Marquis, K.B. Averyt, M. Tignor, and H.L. Miller, Eds., Cambridge University Press, 747-845.

Meng, Q., M. Latif, W. Park, N. S. Keenlyside, V. A. Semenov, and T. Martin, 2011: Twentieth century Walker Circulation change: data analysis and model experiments, *Clim. Dyn.*, **38**, 1757-1773.

Murakami, H., B. Wang, and A. Kitoh, 2011: Future change of western North Pacific typhoons: Projections by a 2-km-mesh global atmospheric model. *J. Climate*, **24**, 1154-1169.

Monahan, A. H., and A. Dai, 2004: The spatial and temporal structure of ENSO nonlinearity. *J. Climate*, **17**, 3026-3036.

NOAA, cited 2012a: Central North Pacific Climatology Overview. [Available online at: http://www.pacificstormsclimatology.org/index.php?page=overview-cnp.]

——, cited 2012b: Western North Pacific Climatology Overview. [Available online at: http://www.pacificstormsclimatology.org/index.php?page=overview-wnp.]

Norton, C. W., P.-S. Chu, and T. A. Schroeder, 2011: Projecting changes in future heavy rainfall events for Oahu, Hawaii: A statistical downscaling approach. *J. Geophys. Res.* **116,** D17110.

Overland, J. E., M. Wang, N. A. Bond, J. E. Walsh, V. M. Kattsov, and W. L. Chapman, 2011: Considerations in the selection of global climate models for regional climate projections: The Arctic as a case study. *J. Climate*, **24,** 1583-1597.

PCMDI, cited 2012: CMIP3 Climate Model Documentation, References, and Links. [Available online at http://www-pcmdi.llnl.gov/ipcc/model_documentation/ipcc_model_documentation.php.]

Power, S. B., and G. Kociuba, 2011: What caused the observed 20th century weakening of the Walker circulation?, *J. Climate.*, **24,** 6501-6514.

Randall, D.A., and Coauthors, 2007: Climate models and their evaluation. *Climate Change 2007: The Physical Basis. Contribution of Working Group I to the Fourth Assessment Report of the Intergovernmental Panel on Climate Change*, Solomon, S., D. Qin, M. Manning, Z. Chen, M.

Marquis, K.B. Averyt, M. Tignor, and H.L. Miller, Eds., Cambridge University Press, 590-662.

Rayner, N., D. Parker, E. Horton, C. Folland, L. Alexander, D. Rowell, E. Kent, and A. Kaplan, 2003: Global analyses of sea surface temperature, sea ice, and night marine air temperature since the late nineteenth century, *J. Geophys. Res.*, **108**, 4407.

Smith, T.M., and R.W. Reynolds, 2004: Improved Extended Reconstruction of SST (1854-1997). *J. Climate*, **17**, 2466-2477.

Smith, T. M., R. W. Reynolds, T. C. Peterson, and J. Lawrimore, 2008: Improvements to NOAA's historical merged land-ocean surface temperature analysis (1880-2006), *J. Climate*, **21**, 2283-2296.

Sohn, B., and S. C. Park, 2010: Strengthened tropical circulations in past three decades inferred from water vapor transport, *J. Geophys. Res.*, **115**, D15112.

Tebaldi, C., J. M. Arblaster, and R. Knutti, 2011: Mapping model agreement on future climate projections. *Geophys. Res. Lett.*, **38,** L23701.

Timm, O. E., H. Diaz, T. Giambelluca, and M. Takahashi, 2011: Projection of changes in the frequency of heavy rain events over Hawai'i based on leading Pacific climate modes, *J. Geophys. Res.*, **116,** D04109.

Tokinaga, H., S. P. Xie, A. Timmermann, S. McGregor, T. Ogata, H. Kubota, and Y. M. Okumura, 2011: Regional Patterns of Tropical Indo-Pacific Climate Change: Evidence of the Walker Circulation Weakening, *J. Climate.*, **25,** 1689-1710.

Van Nguyen, H. V., Y. L. Chen, and F. Fujioka, 2010: Numerical simulations of island effects on airflow and weather during the summer over the Island of Oahu. *Mon. Wea. Rev.*, **138**, 2253-2280.

Vecchi, G., A. Clement, and B. Soden, 2008: Examining the tropical Pacific's response to global warming, *Eos Trans. AGU*, **89,** 81–83.

Vincent, D. G., 1994: The South Pacific convergence zone (SPCZ): A review. *Mon. Weat. Rev.*, **122**, 1949-1970.

Webster, P. J., G. J. Holland, J. A. Curry, and H. R. Chang, 2005: Changes in tropical cyclone number, duration, and intensity in a warming environment. *Science*, **309,** 1844-1846.

Wilby, R. L., and T. Wigley, 1997: Downscaling general circulation model output: a review of methods and limitations. *Prog. Phys. Geog.*, **21,** 530.

WRCC, cited 2012: Climate of Hawai'i. [Available online at http://www.wrcc.dri.edu/narratives/HAWAII.htm.]

Wyant, M.C., 2010: The PreVOCA experiment: modeling the lower troposphere in the Southeast Pacific. *Atmos. Chem. Phys.*,**10**, 4757-4774.

Xie, S.-P., C. Deser, G.A. Vecchi, J. Ma, H. Teng and A.T. Wittenberg, 2010: Global warming pattern formation: Sea surface temperature and rainfall. *J. Climate*, **23**, 966-986.

Young, W. J., 2007: Climate Risk Profile for Samoa. Samoa Meteorology Division. [Available online at: http://pacificdisastermanagement.kemlu.go.id/Documents/Samoa/Climate_Risk_Profile.pdf.]

Yu, B., and F. W. Zwiers, 2010: Changes in equatorial atmospheric zonal circulations in recent decades, *Geophys. Res. Lett.*, **37,** L05701.

Yu, Z.-P., P.-S. Chu, and T.A. Schroeder, 1997: Predictive skills of seasonal to annual rainfall variations in the U.S. affiliated Pacific Islands: Canonical correlation analysis and multivariate principal component regression approaches. *J. Climate*, 10, 2586-2599.

Zhang, Y., Y.-L. Chen, and K. Kodama, 2005a: Validation of the coupled NCEP Mesoscale Spectral Model and an advanced Land Surface Model over the Hawaiian Islands. Part II: A high wind event. *Wea. Forecasting*, **20,** 873-895.

Zhang, Y., Y. -L. Chen, S.-Y. Hong, H.-M. H. Juang, and K. Kodama, 2005b: Validation of the coupled NCEP Mesoscale Spectral Model and an advanced Land Surface Model over the Hawaiian Islands. Part I: Summer trade wind conditions and a heavy rainfall event. *Wea. Forecasting,* **20,** 847-872.

Zhang, C., Y. Wang and K. Hamilton, 2011: Improved representation of boundary layer clouds over the Southeast Pacific in WRF-ARW Using a modified Tiedtke cumulus scheme. *Mon. Wea. Rev.*, **139,** 3489-3513.

Zhang, C., Y. Wang, A. Lauer and K. Hamilton, 2012: Configuration and evaluation of the WRF model for the study of Hawaiian regional climate. *Mon. Wea. Rev.,* **140,** 3259-3277.

# 6. ACKNOWLEDGEMENTS

We acknowledge the modeling groups, the Program for Climate Model Diagnosis and Intercomparison (PCMDI) and the WCRP's Working Group on Coupled Modelling (WGCM) for their roles in making available the WCRP CMIP3 multi-model dataset. Support of this dataset is provided by the Office of Science, U.S. Department of Energy. Analysis of the CMIP3 GCM simulations was provided by Michael Wehner of the Lawrence Berkeley National Laboratory and by Jay Hnilo of the Cooperative Institute for Climate and Satellites (CICS). Document support was provided by Fred Burnett and Clark Lind of TBG Inc. Additional programming and graphical support was provided by Greg Dobson of the University of North Carolina-Asheville.

A partial listing of reports appears below:

NESDIS 102  NOAA Operational Sounding Products From Advanced-TOVS Polar Orbiting Environmental Satellites. Anthony L. Reale, August 2001.

NESDIS 103  GOES-11 Imager and Sounder Radiance and Product Validations for the GOES-11 Science Test. Jaime M. Daniels and Timothy J. Schmit, August 2001.

NESDIS 104  Summary of the NOAA/NESDIS Workshop on Development of a Coordinated Coral Reef Research and Monitoring Program. Jill E. Meyer and H. Lee Dantzler, August 2001.

NESDIS 105  Validation of SSM/I and AMSU Derived Tropical Rainfall Potential (TRaP) During the 2001 Atlantic Hurricane Season. Ralph Ferraro, Paul Pellegrino, Sheldon Kusselson, Michael Turk, and Stan Kidder, August 2002.

NESDIS 106  Calibration of the Advanced Microwave Sounding Unit-A Radiometers for NOAA-N and NOAA-N=. Tsan Mo, September 2002.

NESDIS 107  NOAA Operational Sounding Products for Advanced-TOVS: 2002. Anthony L. Reale, Michael W. Chalfant, Americo S. Allegrino, Franklin H. Tilley, Michael P. Ferguson, and Michael E. Pettey, December 2002.

NESDIS 108  Analytic Formulas for the Aliasing of Sea Level Sampled by a Single Exact-Repeat Altimetric Satellite or a Coordinated Constellation of Satellites. Chang-Kou Tai, November 2002.

NESDIS 109  Description of the System to Nowcast Salinity, Temperature and Sea nettle (*Chrysaora quinquecirrha*) Presence in Chesapeake Bay Using the Curvilinear Hydrodynamics in 3-Dimensions (CH3D) Model. Zhen Li, Thomas F. Gross, and Christopher W. Brown, December 2002.

NESDIS 110  An Algorithm for Correction of Navigation Errors in AMSU-A Data. Seiichiro Kigawa and Michael P. Weinreb, December 2002.

NESDIS 111  An Algorithm for Correction of Lunar Contamination in AMSU-A Data. Seiichiro Kigawa and Tsan Mo, December 2002.

NESDIS 112  Sampling Errors of the Global Mean Sea Level Derived from Topex/Poseidon Altimetry. Chang-Kou Tai and Carl Wagner, December 2002.

NESDIS 113  Proceedings of the International GODAR Review Meeting: Abstracts. Sponsors: Intergovernmental Oceanographic Commission, U.S. National Oceanic and Atmospheric Administration, and the European Community, May 2003.

NESDIS 114  Satellite Rainfall Estimation Over South America: Evaluation of Two Major Events. Daniel A. Vila, Roderick A. Scofield, Robert J. Kuligowski, and J. Clay Davenport, May 2003.

NESDIS 115  Imager and Sounder Radiance and Product Validations for the GOES-12 Science Test. Donald W. Hillger, Timothy J. Schmit, and Jamie M. Daniels, September 2003.

NESDIS 116  Microwave Humidity Sounder Calibration Algorithm. Tsan Mo and Kenneth Jarva, October 2004.

NESDIS 117  Building Profile Plankton Databases for Climate and EcoSystem Research. Sydney Levitus, Satoshi Sato, Catherine Maillard, Nick Mikhailov, Pat Cadwell, Harry Dooley, June 2005.

NESDIS 118  Simultaneous Nadir Overpasses for NOAA-6 to NOAA-17 Satellites from 1980 and 2003 for the Intersatellite Calibration of Radiometers. Changyong Cao, Pubu Ciren, August 2005.

NESDIS 119  Calibration and Validation of NOAA 18 Instruments. Fuzhong Weng and Tsan Mo, December 2005.

NESDIS 120  The NOAA/NESDIS/ORA Windsat Calibration/Validation Collocation Database. Laurence Connor, February 2006.

NESDIS 121  Calibration of the Advanced Microwave Sounding Unit-A Radiometer for METOP-A. Tsan Mo, August 2006.

NESDIS 122    JCSDA Community Radiative Transfer Model (CRTM). Yong Han, Paul van Delst, Quanhua Liu, Fuzhong Weng, Banghua Yan, Russ Treadon, and John Derber, December 2005.

NESDIS 123    Comparing Two Sets of Noisy Measurements. Lawrence E. Flynn, April 2007.

NESDIS 124    Calibration of the Advanced Microwave Sounding Unit-A for NOAA-N'. Tsan Mo, September 2007.

NESDIS 125    The GOES-13 Science Test: Imager and Sounder Radiance and Product Validations. Donald W. Hillger, Timothy J. Schmit, September 2007.

NESDIS 126    A QA/QC Manual of the Cooperative Summary of the Day Processing System. William E. Angel, January 2008.

NESDIS 127    The Easter Freeze of April 2007: A Climatological Perspective and Assessment of Impacts and Services. Ray Wolf, Jay Lawrimore, April 2008.

NESDIS 128    Influence of the ozone and water vapor on the GOES Aerosol and Smoke Product (GASP) retrieval. Hai Zhang, Raymond Hoff, Kevin McCann, Pubu Ciren, Shobha Kondragunta, and Ana Prados, May 2008.

NESDIS 129    Calibration and Validation of NOAA-19 Instruments. Tsan Mo and Fuzhong Weng, editors, July 2009.

NESDIS 130    Calibration of the Advanced Microwave Sounding Unit-A Radiometer for METOP-B. Tsan Mo, August 2010.

NESDIS 131    The GOES-14 Science Test: Imager and Sounder Radiance and Product Validations. Donald W. Hillger and Timothy J. Schmit, August 2010.

NESDIS 132    Assessing Errors in Altimetric and Other Bathymetry Grids. Karen M. Marks and Walter H.F. Smith, January 2011.

NESDIS 133    The NOAA/NESDIS Near Real Time CrIS Channel Selection for Data Assimilation and Retrieval Purposes. Antonia Gambacorta, Chris Barnet, Walter Wolf, Thomas King, Eric Maddy, Murty Divakarla, Mitch Goldberg, April 2011.

NESDIS 134    Report from the Workshop on Continuity of Earth Radiation Budget (CERB) Observations: Post-CERES Requirements. John J. Bates and Xuepeng Zhao, May 2011.

NESDIS 135    Averaging along-track altimeter data between crossover points onto the midpoint gird: Analytic formulas to describe the resolution and aliasing of the filtered results. Chang-Kou Tai, August 2011.

NESDIS 136    Separating the Standing and Net Traveling Spectral Components in the Zonal-Wavenumber and Frequency Spectra to Better Describe Propagating Features in Satellite Altimetry. Chang-Kou Tai, August 2011.

NESDIS 137    Water Vapor Eye Temperature vs. Tropical Cyclone Intensity. Roger B. Weldon, August 2011.

NESDIS 138    Changes in Tropical Cyclone Behavior Related to Changes in the Upper Air Environment. Roger B. Weldon, August 2011.

NESDIS 139    Computing Applications for Satellite Temperature Datasets: A Performance Evaluation of Graphics Processing Units. Timothy F.R. Burgess and Scott F. Heron, December 2011.

NESDIS 140    Microburst Nowcasting Applications of GOES. Kenneth L. Pryor, September 2011.

NESDIS 141    The GOES-15 Science Test: Imager and Sounder Radiance and Product Validations. Donald W. Hillger and Timothy J. Schmit, November 2011.

# NOAA SCIENTIFIC AND TECHNICAL PUBLICATIONS

*The National Oceanic and Atmospheric Administration* was established as part of the Department of Commerce on October 3, 1970. The mission responsibilities of NOAA are to assess the socioeconomic impact of natural and technological changes in the environment and to monitor and predict the state of the solid Earth, the oceans and their living resources, the atmosphere, and the space environment of the Earth.

The major components of NOAA regularly produce various types of scientific and technical information in the following types of publications

**PROFESSIONAL PAPERS** – Important definitive research results, major techniques, and special investigations.

**CONTRACT AND GRANT REPORTS** – Reports prepared by contractors or grantees under NOAA sponsorship.

**ATLAS** – Presentation of analyzed data generally in the form of maps showing distribution of rainfall, chemical and physical conditions of oceans and atmosphere, distribution of fishes and marine mammals, ionospheric conditions, etc.

**TECHNICAL SERVICE PUBLICATIONS** – Reports containing data, observations, instructions, etc. A partial listing includes data serials; prediction and outlook periodicals; technical manuals, training papers, planning reports, and information serials; and miscellaneous technical publications.

**TECHNICAL REPORTS** – Journal quality with extensive details, mathematical developments, or data listings.

**TECHNICAL MEMORANDUMS** – Reports of preliminary, partial, or negative research or technology results, interim instructions, and the like.

**U.S. DEPARTMENT OF COMMERCE**
National Oceanic and Atmospheric Administration
National Environmental Satellite, Data, and Information Service
Washington, D.C. 20233

www.ingramcontent.com/pod-product-compliance
Lightning Source LLC
Chambersburg PA
CBHW060527210526
45168CB00024B/3368